Lean Methodology

A Guide to Lean Six Sigma, Agile Project Management, Scrum and Kanban for Beginners

© **Copyright 2019**

All Rights Reserved. No part of this book may be reproduced in any form without permission in writing from the author. Reviewers may quote brief passages in reviews.

Disclaimer: No part of this publication may be reproduced or transmitted in any form or by any means, mechanical or electronic, including photocopying or recording, or by any information storage and retrieval system, or transmitted by email without permission in writing from the publisher.

While all attempts have been made to verify the information provided in this publication, neither the author nor the publisher assumes any responsibility for errors, omissions or contrary interpretations of the subject matter herein.

This book is for entertainment purposes only. The views expressed are those of the author alone, and should not be taken as expert instruction or commands. The reader is responsible for his or her own actions.

Adherence to all applicable laws and regulations, including international, federal, state and local laws governing professional licensing, business practices, advertising and all other aspects of doing business in the US, Canada, UK or any other jurisdiction is the sole responsibility of the purchaser or reader.

Neither the author nor the publisher assumes any responsibility or liability whatsoever on the behalf of the purchaser or reader of these materials. Any perceived slight of any individual or organization is purely unintentional.

Contents

INTRODUCTION 1
PART ONE: PROJECT MANAGEMENT BASICS 3
CHAPTER ONE: PROJECT MANAGEMENT METHODOLOGIES 4
    WHAT IS PROJECT MANAGEMENT? 4
    PROJECT MANAGEMENT METHODOLOGIES 5
        Agile 6
        Scrum 8
        Events 8
        Scrum Artifacts 9
        Kanban 9
        Lean 11
        Waterfall 13
        Six Sigma 13
        PMI/PMBOK 15
CHAPTER TWO: THE PROJECT MANAGER ROLE 17
    WHO IS A PROJECT MANAGER? 18

ROLES AND RESPONSIBILITIES OF PROJECT MANAGERS 18

   Resource and Activity Planning *18*

   Motivating and Organizing the Team *19*

   Time Management *19*

   Estimating Cost and Developing the Budget *19*

   Ensuring Customer Satisfaction *19*

   Identifying and Managing Project Risk *20*

   Monitoring Progress *20*

   Managing Necessary Documentation and Reports *20*

DO YOU NEED A PROJECT MANAGER? 20

CHAPTER THREE: CHOOSING THE RIGHT METHODOLOGY 22

BENEFITS OF CHOOSING AN ORGANIZATIONAL PM METHOD 22

CHOOSING THE RIGHT METHOD 23

SOFTWARE 24

   Gantt Charts *25*

   Kanban Boards *25*

   Calendars *25*

   Mobile Accessibility *25*

   Cross-Project Summary Views *26*

PART TWO: LEAN PROJECT MANAGEMENT 27

CHAPTER FOUR: WHAT IS LEAN? 28

LEAN MANAGEMENT FOR PRODUCTION AND SERVICES 29

LEAN BUSINESS PRINCIPLES 29

   Perfection *29*

   Value Identification *30*

   Value Stream Mapping *30*

   Flow *30*

   Pull *31*

LEAN TOOLS 31

  Takt Time *31*

  Continuous Flow *32*

  Standardized Work *32*

  Kanban and Pull Systems *32*

  Cellular Manufacturing *32*

  The Why's *32*

  Level the Workload *33*

  Problem Solving *33*

CHAPTER FIVE: START A LEAN PROJECT **34**

 INITIATION 34

 PLANNING 34

 EXECUTION 35

  Flow *35*

  Inventory *35*

  Kaizen *36*

 MONITORING & CONTROL 36

 CLOSING 36

CHAPTER SIX: LEAN SIX SIGMA **38**

 SIMILARITIES OF LEAN AND SIX SIGMA 38

 WHAT IS THE DIFFERENCE BETWEEN LEAN AND SIX SIGMA? 40

 LEAN SIX SIGMA PRINCIPLES 41

  Addressing a Real-World Problem *41*

  A Team Accomplishes Analysis *42*

  The Analysis is Focused on a Process *42*

  The Analysis is Based on Data *43*

  Understand the Impact of the Process Sigma *44*

  The Solution Addresses the Real Root Cause(s) *45*

 BENEFITS OF LEAN SIX SIGMA 46

  Organizational Benefits *46*

- *Personal Benefits* 48
- INDUSTRIES AND FUNCTIONS USING LEAN SIX SIGMA 49

CHAPTER SEVEN: LEAN STARTUP 52

- PRINCIPLES OF LEAN STARTUP 53
  - *Validated Learning* 53
  - *Entrepreneurs are Everywhere* 53
  - *Controlled Use and Deployment of Resources* 53
  - *Entrepreneurship is Management* 54
  - *Build-Measure-Learn* 54
- MODELS AND METHODOLOGIES 54
  - *Build-Measure-Learn* 54
  - *Minimal Viable Product (MVP)* 55
  - *Validated Learning* 56
  - *Innovation Accounting* 57
  - *Persevere or Pivot* 57
  - *Small Batches* 59
  - *Andon Cord* 60
  - *Continuous Deployment* 60
  - *Kanban* 61
  - *The Five Whys* 62

CHAPTER EIGHT: LEAN ENTERPRISE 64

- HOW TO START TO CREATE A LEAN ENTERPRISE 64
- THE DIFFICULTIES OF THE FLOW OF INFORMATION 65
- SIPOC TO STANDARDIZE THE INFORMATION FLOW 66

CHAPTER NINE: LEAN TEAMS 68

- DEVELOPING A LEAN TEAM 68
- FORMING A TEAM 68
- EMPOWERING THE TEAM 69
- LEAN TEAM HIERARCHY 70

- IMPLEMENTING A LEAN TEAM 70
  - *Kaizen* 71
  - *Five Whys* 71
  - *PDCA* 71
- HOW TO BUILD A LEAN TEAM 71
  - *Start Small* 72
  - *Make the Team Cross-Functional* 72
  - *Never Over-Rely on Team Players* 73
  - *Train People to Be Team Smart* 73
  - *Creating a Pro-Risk Environment* 74
  - *Understanding the Needs of the Team* 75
  - *Measure to Learn and Improve the Team* 75

CHAPTER TEN: WHAT IS LEAN ANALYTICS? 77

- WHAT KIND OF BUSINESS ARE YOU? 77
  - *Empathy* 78
  - *Stickiness* 78
  - *Virality* 78
  - *Revenue* 78
  - *Scale* 78
- HOW TO APPLY LEAN ANALYTICS 78
  - *Trigger* 79
  - *Action* 79
  - *Variable Reward* 79
  - *Investment* 80
- HOW TO USE THE LEAN ANALYTICS CANVAS 80

PART THREE: AGILE PROJECT MANAGEMENT 81

CHAPTER ELEVEN: WHAT IS THE AGILE FRAMEWORK? 82

- WHICH FRAMEWORK IS BEST? 83
- SCRUM, EXTREME PROGRAMMING AND KANBAN 83

- The Agile Manifesto 84
- Agile Principles 85
- Platinum Principles 87
  - *Visualize Instead of Writing* 87
  - *Think and Act as a Team* 88
  - *Avoiding Formality* 88

CHAPTER TWELVE: START AN AGILE PROJECT 90

- Understanding Agile Project Management 90
  - *Understand the Problem* 91
  - *Assemble the Right Team* 91
  - *Brainstorm* 91
  - *Build an Initial Prototype* 91
  - *Decide the Boundaries* 92
  - *Plan the Major Milestones Using A Roadmap* 92
  - *Plan Sprints* 92
  - *Check In Every Day* 93
  - *Review the Sprint* 93
  - *Plan the Next Sprint* 93
  - *Completion and Release* 94

CHAPTER THIRTEEN: AGILE VERSUS SCRUM VERSUS KANBAN ... 95

- Differences Between Agile, Kanban and Scrum 95
- Differences Between Scrum and Kanban 96
  - *Product Owner* 96
  - *Scrum Master* 97
  - *Team Members* 97
- Agile Pros and Cons 97
- Scrum Pros and Cons 98
- Kanban Pros and Cons 98
- Which One Should You Choose? 99

CHAPTER FOURTEEN: STEP-BY-STEP SCRUM 100
BASICS OF SCRUM 101
THE ROADMAP TO VALUE 102
A SIMPLE OVERVIEW 103
Product Backlog 103
Teams 104
GOVERNANCE 105
Product Owner 105
Development Team 105
Scrum Master 106
SCRUM FRAMEWORK 106
Artifacts 107
Roles 107
Events 107
FEEDBACK 108
STEPS TO FOLLOW 109
Define the Scrum Team 109
Define the Sprint Length 109
Appoint the Scrum Master 109
Appoint the Product Owner 110
Create the Initial Backlog 110
Plan the Start of The First Sprint 111
Close the Current Sprint and Start the Next One 111

CHAPTER FIFTEEN: CREATE A KANBAN PROJECT 112
A BRIEF HISTORY 112
WHAT IS THE KANBAN METHOD? 113
KANBAN CHANGE MANAGEMENT PRINCIPLES 113
FOUNDATIONAL PRINCIPLES 114
Always start with what you are doing now 114

- *Pursue Evolutionary and Incremental Change* 114
- *Respect Current Roles, Responsibilities, and Designations* 114
- *Encourage Acts of Leadership* 114
- CORE PRACTICES 114
  - *Visualize the Workflow* 114
  - *Limit or Reduce Work-in-Progress* 115
  - *Manage Flow* 115
  - *Make Policies Explicit* 116
  - *Feedback Loops* 117
  - *Improve and Evolve Collaboratively and Experimentally* 117
- IMPLEMENTING KANBAN 118
  - *Step One: Visualization Of Workflow* 118
  - *Step Two: Limit the Amount of WIP* 118
  - *Step Three: Switch to Explicit Policies* 119
  - *Step Four: Measure and Manage the Workflow* 119
  - *Step Five: Using Scientific Methods for Optimization* 119

CONCLUSION **120**

RESOURCES **122**

Introduction

You may have come across the phrase, "Life is what happens to you while you are busy making other plans." This applies to a business, too. A business should learn to change or adapt to change since things never go according to plan. The same can be said about the different teams in an organization. In this era of technology, many businesses are shifting toward artificial intelligence and automation. A machine can perform different processes. For example, if you own a company that sells cassettes, you know you cannot make sales because the world has moved on from cassettes to storing music on mobile devices. The same can be said about some processes in an organization.

Many processes in organizations become redundant over a period, and they should be removed from the organization. It is only when the business can do this that it can apply one of the new-age project management methodologies. Throughout this book, you will gather information about the basics of project management. You will learn more about what the role of a project manager is and how you should choose the right methodology for your project.

Numerous project management methodologies have been developed to improve businesses and processes. You must understand these methodologies to ensure that you know what method to choose.

Having said that, you cannot expect a method that worked in one company to work in yours. Each team is different, and the same goes for your team, too. When you choose a project management methodology for your team or business, remember that you need to tweak the method so that it works well for you and your team.

This book also sheds some light on methodologies like Lean and Agile. You will gather information about lean project management and learn more about the different components of lean project management. You will gather information on how to start a lean project and also learn more about lean six sigma. The book also details the agile framework and addresses the different methodologies that come under the umbrella of the agile framework. It is only when you fully understand these concepts that you can work better with your teams.

PART ONE: Project Management Basics

Chapter One: Project Management Methodologies

What Is Project Management?

A project is often a temporary undertaking for creating a specific service, product, or obtaining the desired result. The scope and resources involved for completing a project are always defined along with the time taken for its completion. Apart from this, the dates on which the project will start and end are also identified. In this sense, the project is a temporary endeavor. A project is different from all the other routine operations taking place in an organization. It has a unique purpose and includes specific operations that are designed with the sole purpose of attaining unique goals. All the members present in a project work towards attaining a singular goal. It means that even those employees who don't usually work together are clubbed together for attaining these goals. All the members of a project team don't necessarily have to be from the same organization and, at times, are from different organizations across various geographies. Creation of software to streamline business operations, constructing a new building, working together after a natural calamity, and expanding into new markets are all examples of projects.

Given the complexity of a project, it is quintessential that they are thoroughly managed for delivering results on time while sticking to the budget forecast. Therefore, project management is the application of different skills, tools, knowledge, and techniques associated with all project-related activities for obtaining the project goals. Project management has always been a part of any organization or enterprise. However, it was only during the mid-twentieth century that it was recognized as a distinct profession. There are five processes involved in any project management, and they are initiation, planning, execution, monitoring and control, and closing.

Integration, scope, cost, time, quality, human resources, procurement, communications, stakeholder management, and risk management are all the different areas of knowledge involved in project management. Management is often concerned with all these areas. However, project management does lend a unique focus based on the goals, resources, and schedules associated with each project.

Project Management Methodologies

Project management methodology refers to the different practices, techniques, rules, and procedures used by a project manager. Examples of project management methodologies include Six Sigma, Kanban, Agile, and so on. These methodologies encompass different processes that offer guidance to project managers through different stages of a project and enable them to complete various tasks involved. They essentially help to manage and maximize the resources as well as the time available.

Selecting the right project management methodology is an important duty. Project managers understand its importance since it is essential for getting the work done effectively and efficiently. There are different types of project management methodologies, and no one size fits all solution is available. In fact, there is no such thing as the right methodology that the manager can choose—it principally means that you cannot opt for one methodology for all the different projects you undertake. Depending on the scope and requirements of the project, the management methodology you use will change. In this section, you will learn about the most popular project management methodologies.

Agile

One of the popular project management methodologies today is Agile. It is well-suited for all such projects that are incremental and iterative. It consists of processes wherein demands as well as solutions are obtained via collaborative efforts of cross-functional teams, self-organization, and the customers. Initially, it was created with the aim of software development. Agile was established to cope with the shortcomings of the Waterfall method. The processes of this were incapable of meeting the various demands of an extremely competitive and constantly changing environment of the software industry.

The values, as well as the principles of Agile project management methodology, are obtained from the Agile manifesto. Thirteen industry leaders formalized this declaration in 2001. Agile project management aimed to come up with different ways to develop software by providing a measurable and clear structure. This structure supports iterative development, recognition of changes, and provides a framework for team collaborations. There are twelve key principles, along with four fundamental values that form the basis of Agile project management methodology.

The values of this methodology are as follows:

- Emphasis is placed on interactions as well as individuals, more than on the tools and processes involved.

- Priority is given to customer collaboration instead of contract negotiation.

- Instead of comprehensive documentation, this framework enables the working of the software.

- Instead of blindly following a plan, Agile methodology fosters better responses to change.
- The principles of this management methodology are as follows:

1. Customer satisfaction is ensured via continuous and early software delivery.

2. The requirements of a project keep changing throughout the developmental process. Agile management helps accommodate all of these requirements.

3. Continuous delivery of working software.

4. It provides a basis for collaborative efforts between developers as well as the business stakeholders during the entire duration of the project.

5. It helps foster trust, motivation, and support amongst all those who are involved.

6. It increases the scope for face-to-face interactions.

7. The primary measure of progress used in this methodology is the working software.

8. It creates a basis for the consistent and continuous pace of development.

9. The simplicity it offers is unlike other methodologies.

10. Great attention to design and technical details.

11. The self-organization of teams prompts the development of designs, requirements, and architectures.

12. It provides scope for regular reviews to improve effectiveness.

The adaptiveness of Agile methodology comes in handy, especially while working on complex projects. It uses different deliverables to measure and track progress to help with project completion and product development. The six deliverables used by Agile include release plan, vision statement, product backlog, sprint backlog, increment, and product roadmap. Because of these features, the main emphasis of this methodology is flexibility, frequent improvements, collaboration, and delivering high-quality outputs.

Scrum

The five values of Scrum are based on include respect, commitment, openness, focus, and courage. The primary goal of Scrum is to help with the development, delivery, and the sustenance of complex products. All this is attained via collaboration, iterative progress, and accountability. The main factor that differentiates Scrum from Agile project management is that it functions using specific events, roles, and artifacts. These three factors will now be looked at in detail.

Team Roles

There are three team roles available in Scrum management: product owners, the development team, and the Scrum Master. The product owner, as the name suggests, represents all the stakeholders as well as potential customers. The development team comprises of different professionals who help and deliver the product, such as programmers, designers, and developers. The third category includes an organized manager who helps understand and execute how Scrum has to be followed.

Events

There are five events included in Scrum: Sprint, Sprint Planning, Daily Scrum, Sprint review, and sprint retrospective. Sprint is the term used for the iterative time boxes within which a goal has to be attained. The timeframe for every individual box is one calendar month, and it cannot exceed this timeframe. Also, the same will be consistent throughout the development process. Whenever the Scrum team gets together, usually at the beginning of the Sprint, it is known as Sprint planning. During this meeting, the team plans for the Sprint. Daily Scrum refers to fifteen minutes of the daily meeting, which will be held at the same time consistently. In this meeting, a review of the previous day's progress is discussed, along with any expectations for the subsequent day. At the end of every Sprint, an informal meeting is held wherein the Scrum team presents their progress and achievements to the stakeholders. This meeting is known as a Sprint review, and it helps the Scrum team obtain feedback from the stakeholders. The final event is the Sprint retrospective. This is a meeting when the Scrum team reviews the proceedings of the previous Sprint and comes up with ways in which improvements can be implemented for the upcoming Sprint.

Scrum Artifacts

There are two Scrum artifacts you must familiarize yourself with: the product backlog and the Sprint backlog. The product backlog is taken care of by the product owner. All the requirements desire from the viable product will be listed according to their priorities in the product backlog. It includes information about the different features, requirements, functions, any enhancements, and other fixes that showcase any changes required to be made to the product.

The Sprint backlog contains a detailed list of tasks as well as requirements that need to be attained during the subsequent Sprint. At times, a Scrum task board is also used for visualizing the progress of the tasks in the ongoing Sprint. Apart from this, any changes which are required to be made to the product are also listed on the Scrum task board using three different columns based on the status: to do, doing, and done.

Kanban

Kanban is a simple yet popular Agile methodology that focuses on the early release of products via collaboration between self-managing teams. It is quite similar to Scrum. The concept of Kanban was developed in the 1940s for optimizing the production line in the Toyota factories. Kanban offers the visual representation of the production process to deliver high-quality results by showing the workflow to determine any bottlenecks. The sooner the bottlenecks are identified in the workflow process, the easier it is to tackle them. Since any potential bottlenecks are identified during the early stages of the development process, it increases the overall efficiency. Six practices form the basis of Kanban:

- Visualization of the workflow

- Limiting the "work in progress" tasks

- Management of the workflow

- Explicitly stating all the policies

- Optimization and utilization of feedback loops

- Collaboration and experimental evolution of the processes

The visual cues used in a Kanban help identify the different stages of a developmental process. There are two key processes used in Kanban: the Kanban board and the Kanban cards. Apart from this, Kanban swimlanes can be used for a little extra organization.

The Kanban board helps visualize the developmental process. Either it can be a physical board, like sticky notes and a whiteboard, or a digital board. Digital boards are provided by different online project management tools like Zenkit. A Kanban card represents the item or task in progress. It helps communicate the workflow and progress to the team. It also determines vital information like the timeframe of the project, any upcoming deadlines, and the status of work. The Kanban swimlanes flow horizontally and provide for further classification of items via categorization. These swimlanes improve the overview of the work going on.

There are no specific rules you need to follow while using the Kanban project methodology. Use a Kanban board to visually represent the different stages of development starting from when the ideas were produced to ongoing work and finally the completion of work. Usually, the Kanban board is divided into three columns. The first column consists of all the tasks that need to be completed. The second column shows the work in progress, and the final one shows the completion of tasks. Kanban is quite popular in the software development industry. The flexibility it offers makes it the ideal choice in other industries, too. If a project requires any improvements during the development process, then Kanban will come in handy.

Lean

The lean management methodology helps minimize wastage while maximizing customer value. It is based on a simple principle that helps optimize the value a customer derives by reducing the use of resources. This management methodology was created in the Japanese manufacturing industry. The quality of the final output improves when waste is eliminated, and the costs, along with the production time, are reduced. Muda, Mura, and Muri are the three types of wastes identified by Lean methodology.

Any process that doesn't add any value to the project must be removed. Muda refers to this type of waste. Wastage can be of tangible and intangible resources. For instance, it can be a waste of physical resources or an intangible yet precious resource like time available. There are seven wastes identified by the Lean methodology:

- The transport or movement of a product between the locations and operations.

- Inventory or the work in progress along with any inventory of finished products and raw materials an organization maintains.

- Any physical movement made by a human being or a machine whenever an operation is conducted.

- Merely waiting for a product to arrive, for a machine to complete its task, or any other reason is a waste of time.

- Producing more than the demand or what the customer has ordered.

- Conducting unnecessary processes or operations that exceed the customer's requirements.

- Any defects in the form of product reject or rework required within the production process.

The removal of any variances in the workflow process during the scheduling and operational levels is known as Mura. The elimination of such variances helps ensure that everything goes along smoothly. For instance, if the editor of a magazine spends a long time editing an article, then the creative team will have less time for creating the spread before publishing the magazine. It might also mean that they won't have sufficient time to come up with the best possible spread because of the upcoming deadlines. Therefore, by reducing the time spent on editing, every department gets sufficient time to work on their aspect of the article while adhering to the deadline.

To ensure that nothing slows down during the production process, the team should remove some overloads. Muri is all about doing this. At times, business owners, as well as managers, tend to impose unnecessary stress on their processes and employees because of other factors like poor organizational structure, complicated working structure, or even using the wrong management tools. By reducing this overload, the production process can be streamlined, and the efficiency of the employees, along with their output, can be optimized.

Lean management places emphasis on following certain principles instead of implementing specific processes. There are five basic principles upon which the system is created:

- Specification of value by the final customer

- Identification of different steps in the value stream

- Ensuring a continuous product flow

- Enabling customers to pull value from the subsequent upstream activities

- Managing the work process while eliminating any unnecessary processes, steps, or activities.

Waterfall

The waterfall is a conventional project management methodology. It consists of linear and sequential design where progress flows downwards in a specific direction, almost like a waterfall. This technique first came up in the construction and manufacturing industries. It doesn't offer the flexibility of the other methods during the earlier stages of the development process, especially in the aspects related to design. Winston W. Royce came up with the Waterfall methodology in 1970; however, he didn't use the term "Waterfall." This technique essentially states that you cannot move from one phase of development to the next one until the current state is completed. There are six stages of development, and they are requirements of the system and the software, analysis, design, coding, testing, and operations.

This methodology emphasizes the importance of documentation throughout the process of development. It is based on the premise that even if one worker leaves during a development process, his or her replacement can easily start where the previous one left by using the information available in the different documents. It helps prevent any breaks in the flow of production, even if the workers change. Before Agile was introduced, the Waterfall management methodology was used in software development. However, the non-adaptive and rigid design constraints implemented by this methodology was quite challenging. Apart from this, the lack of a development process to obtain customer feedback, coupled with delays in the testing period, paved the way for newer management methodologies.

Six Sigma

The Six Sigma project management methodology was developed in 1986 by the engineers working at Motorola. The goal of this method is improving the quality of the final product by effectively reducing the errors in a process via identification of those things that aren't working and then eliminating them from the process altogether. Empirical and statistical quality management methods combined with the expertise of specialists in these methods are used in the Six Sigma project management methodology. Six Sigma green belts and six Sigma black belts are the two major methodologies within the Six Sigma, and they are both supervised by the Six Sigma Master Black Belts. The first methodology is DMAIC, and the second one is DMADV. The former is used to improve the efficiency of business processes, while the latter is used to create new processes, services, or products.

DMAIC stands for:

- Defining the problem along with the project goals

- Measuring the different aspects of the ongoing processes

- Analyzing all the data to identify any defects in a process

- Improving the efficiency of the project

- Controlling the way that the process will be executed in the future

DMADV stands for:

- Defining the project goals

- Measuring the critical components of a process along with the capabilities of the product

- Analyzing the data and developing different designs for a process and selecting the best from the lot

- Designing and testing the details of a process

- Verifying the design process by running it through simulations along with a pilot program and then handing the reins of the process to the client

Another method of Lean Six Sigma methodology is used to improve the performance of the team by strategically eliminating waste while reducing any variations in the process.

PMI/PMBOK

PMI or Project Management Institute is a non-profit membership association for project management certifications and standards of the organization. The PMI designed the PMBOK. Essentially, it isn't much of a methodology; instead, it is a set of guiding details for a set of standards that influence project management. Project Management Body of Knowledge or PMBOK is a set of standard guidelines and terminology for project management. According to the PMBOK, there are five process groups in every project:

Initiating

This is the stage where the group should define the start of a new phase in an existing project or the start of a new phase.

Planning

In the planning stage, the group should define the scope and objective of the project. Additionally, the group should also define how these objectives would be met.

Executing

In this phase, the group should work towards completing the steps or the processes defined in the plan.

Monitoring and Controlling

In this stage, the project manager will need to track, regulate, and review the performance and progress.

Closing

This is the final stage of the project, where the group will conclude the activities performed across all the process groups, and close the phase or project.

PMBOK also includes the best conventions, techniques, and practices that every team should adhere to. These guidelines are also updated regularly to ensure that they are up-to-date with the latest project management practices. In 2017, the PMBOK released its sixth edition.

When you begin working on a new project or phase in an existing project, you may want to apply many of the methodologies that were discussed in this chapter. This chapter only provides a guideline that you can use to help you select the right methodology for your project. Once you understand these methodologies better, you can perform further research to help you find the best match, and pair this up with a project management tool, like Zenkit.

Chapter Two: The Project Manager Role

Scott Berkun, the author of *Making Things Happen: Mastering Project Management*, stated that a project manager plays the role of a doctor in a team. A doctor leads the trauma team and chooses the procedure the team should perform on the patient. If there is nobody who can handle project management issues in the right way, the project can get into trouble. When do you think the role of a project manager came into existence? Microsoft was working on an ambitious project in the late 1980s, and this project had run into a problem. There were too many teams involved in the project, and they did not know how to coordinate with each other.

So, Microsoft decided to come up with an ingenious solution. They selected one person to take charge of the project. This person had the authority to coordinate the project and organize it. When Microsoft appointed a leader, the team could deliver the project smoothly, and the teams were happy with the dynamics. The result of this strategy was Excel. Microsoft eventually began to appoint someone to manage the project, and thus the role of a project manager was born.

Who Is a Project Manager?

A good project manager is someone with an excellent entrepreneurial mindset. This mindset gives them the necessary skills to not only manage the project but also manage the teams. A project manager can help the team meet the deadline and deliver the result. The success or failure of a project rests only on the project manager, and they are responsible for the result.

A project manager has the required information and knowledge to manage the project. They will need to understand the tasks that they assign to the team members and the technical know-how to ensure that the project moves forward. The latter skill does not only give the project manager the ability to communicate ideas to every person involved in the project but also helps them win the respect of the stakeholders and team members. Since a project manager will influence any of the decisions made in a project more than anybody else in the company, their task is to use all the knowledge they have to win their team members' respect.

Roles and Responsibilities of Project Managers

Resource and Activity Planning

The team and the project manager need to plan the project so they can meet the deadlines. Many projects fail because there was no planning. A good project manager will define the scope of the project and then determine if the resources are available. A good project manager will also know how to realistically set the time estimates and also evaluate the capabilities of the team. The project manager should then create a plan to complete the project and monitor the progress. A project is unpredictable, so a good project manager should know how to make the necessary adjustments to complete the project.

Motivating and Organizing the Team

A good project manager does not demotivate the team by giving them long checklists and elaborate spreadsheets to understand the progress of the project. Instead, they will always focus on their teams and put them in the center. They will work on a plan to enable their teams to meet their deadlines and reach their potential. They will work on removing bureaucracy and steer the team towards the goal.

Time Management

A client will always judge the success or failure of a project based on whether the team delivered the project on time or not. Therefore, it is important to understand that the timeline is non-negotiable. A good project manager knows how to set a realistic deadline for every task and also speak to the team regularly to understand the progress. They also know how to do the following:

- Maintain a schedule

- Define activity

- Estimate the duration of activity

- Sequence activity

- Develop a schedule

Estimating Cost and Developing the Budget

A good project manager knows how to stick to the budget allotted to the project. A project will be a failure if it goes over the budget, even if it is delivered on time and meets the expectations of the client. A good project manager will review the budget frequently and plan the tasks ahead to ensure that the team does not overrun the budget.

Ensuring Customer Satisfaction

A project is only successful when the customer is happy with the result. Every project manager is required to work towards avoiding any unwanted surprises, minimize uncertainty, and include their customers or clients in the project as much as possible. A good project manager is aware that they must communicate effectively with the customers and the business, and give them an update.

Identifying and Managing Project Risk

Regardless of how big or small a project, there will be obstacles, pitfalls, and hurdles that were not included as a part of the plan. A project manager will know how to identify, evaluate, and measure a potential risk before the project begins. They also make a note of the different ways to overcome that pitfall.

Monitoring Progress

During the start of the project, a project manager and his or her team should have a clear vision of the project. They will also hope to produce the desired output or result as per the timelines. This is, however, not always possible since the team will face some obstacles along the way. If things do not go according to the project plan, the project manager will need to monitor the progress, team performance, expenditure, and analyze that data to take corrective measures wherever required.

Managing Necessary Documentation and Reports

Experienced project managers know they need to maintain proper documentation and final reports for the project, and they know how to maintain this. A good project manager can present a comprehensive document that talks about how the team met all the requirements. This document should also provide some information on the history of the project, the team members involved, what tasks were done, and what can be done better in the future.

Do you need a Project Manager?

Regardless of how demanding or large a project is, you will need someone who can consistently and reliably maintain productivity and efficiency. Research shows that high-performing teams and organizations always have a project manager, and this is a profession that is consistently in demand. A project manager is indispensable to a successful project or business, and every business owner needs a leader who has the right skills, vision, and know-how to face any big challenges. These leaders are the only ones who can complete a project successfully and meet the timelines.

A project manager is an integral part of every organization, regardless of whether it is a small agency or a multinational company. The former only needs to hire one project manager, while the latter may need to hire many project managers or a highly-specialized project manager.

Chapter Three: Choosing the Right Methodology

You must choose the right project methodology to ensure that your team successfully delivers the project on budget and time. From Agile to waterfall to Kanban, there are numerous project management methodologies that you can choose from to maximize success.

This chapter will look at the different steps involved in choosing the right project management method based on the needs of the project and team. You must also ensure that the methodology will benefit your team and the project.

Benefits of Choosing an Organizational PM Method

One of the benefits is that these methods provide your team with guidelines, which will allow them to easily handle and establish different tasks in a project, including the budget, resources, timelines, stakeholders, and team members. There is no right way to handle a project. That being said, if you and your team assess the types of projects you work on along with the different components involved, you could make the right decision regarding the methodology. You can choose the one that will match the needs of your project.

It is difficult to choose the right approach to choosing a project management method since there are so many methods available to choose from. When you keep the project, objectives, and target goals in mind, you can choose the method and adapt it to cater to the needs of your project. Additionally, when you look at the pros and cons of different methods, you can choose numerous project management tools that will meet the requirements of your project.

Choosing the Right Method

You should choose the project management method based on the type of process or project that you work on. It is critical to choose the right project management based on some criteria since there are numerous methodologies or frameworks that you can choose from. Some criteria include:

- Scalability of project

- Project focus; for example, the objective or the different activities

- Allotted budget

- Specialization of roles

- Flexibility of timeline

- Stakeholder and customer involvement

- Industry

- Complexity of projects

- The different teams and the number of employees working on this project

- Resistance to change

- Maintaining an inventory of resources to identify those that are available and those that are needed

- Structure of the organization

- Selecting the dates for when the team starts and ends the project

Once you look at these components, you should identify the method that works best for you. Some of the steps included in this phase are:

1. Since you have identified the goal or the objective of the project, you should look at the different variables that impact the tasks in the project. Weigh these variables against the objective or the goal of the project

2. Identify the criteria on which the methodology will have an impact

3. Assess the different methodologies and choose the method that is most relevant to the project

4. Identify the advantages and disadvantages of every methodology and weigh them against each other

5. Examine the methodology that will bring the most efficiency and success to your project, and also identify the project that will add more risk

6. Collaborate with the team and weigh the decision

7. Always document the method that you choose to implement

8. Apply the methodology, and monitor the steps for the success and progress of that methodology

Software

When you have chosen the right method for your project, you must identify the right tool or software to implement that method. There are numerous options out there that will make it difficult for you to choose the right tool or software since there are many variables that you will need to consider. The following are some functionalities to consider when you need to choose the right project management software for your project.

Gantt Charts

Project managers and businesses use Gantt charts to look at the amount of work done. These charts depict the tasks completed in the form of a series of horizontal lines on the chart. You can also use this chart to represent the targets hit during specific times in the project. This view will help you look at the time that the team took to complete a task in comparison to the time that the team was supposed to take as per the plan. This functionality is often used when the waterfall project management method is applied to a project since the team should adhere to a strict timeline.

Kanban Boards

Kanban boards are workflow visualization tools that allow project managers and teams to optimize the workflow. They use these boards to see how the project is progressing. You can use a Kanban board to communicate progress reports, status updates, project issues, and more, thereby offering your team full visibility of the project. The Kanban board is often paired with the Kanban methodology, where the teams are required to focus only on working on the tasks and deliver the tasks and projects over a fluid timeline. These boards are also used in the Lean project management method to allocate tasks to team members and also manage the inventory and resources.

Calendars

You can use a calendar to show the stakeholders and the business a detailed view of the project timelines. You can also show them the dates of completion. This gives the team members, stakeholders, businesses, and customers complete visibility of the status of the project. This view is often used in project methodologies like Scrum and XPM (Extreme Programming).

Mobile Accessibility

When customers and team members can view the status of the project—including the tasks completed, the progress, changes in the timeline, and more—on their mobile device, it is considered important in the overall success of the project. It is easier to use this tool when you use a project management methodology that is fluid. You can also use it when the project has different parts with some dependencies, like Agile.

Cross-Project Summary Views

A cross-project view will give the departments and teams involved in the project a full insight into the development of the project. This feature makes it easier for you to share real-time status updates, dashboards, and roll-up checkpoint details that will give customers, stakeholders, and the team members an update throughout the project. It is best to use this feature with methodologies that require some updates due to the fluid-structure, like Scrum, Agile, and XPM. Having said that, it could be used for other methodologies, too.

PART TWO: Lean Project Management

Chapter Four: What Is Lean?

As mentioned earlier, one of the core principles of Lean is to maximize customer value and minimize waste. In simple words, lean helps to provide customers with a product of greater value using fewer resources. Lean thinking will make the organization focus on those processes that require a change. The management will learn to focus on how the flow of products and services can be optimized horizontally across assets, technology, and departments to its customers instead of focusing on separate technology, vertical departments, and assets.

The management will also have to look at how the processes that add waste to the project should be removed along the value stream. The business needs to look at how the processes should be optimized to reduce human effort, human time, capital, and space. This will help to reduce the cost incurred to finish the product or service. These companies find it easier to meet the changing needs of their customers with high-quality products, great variety, faster delivery, and low cost. Information management will also become more accurate and straightforward.

Lean Management for Production and Services

Most people believe that lean management only works in the manufacturing department, and this is not true. The Lean management methodology can be applied to different types of processes and businesses. It is not a method that is used only to reduce the cost. This methodology gives businesses a set of principles to improve the way the business thinks or acts. Businesses in different industries and sectors, including healthcare and government, have begun to use lean principles to change the way the processes function in the business. Companies that move away from traditional practices and approaches to a lean way of thinking are undergoing a lean transformation. James Womack, Ph.D., who led a research team at an International Motor Vehicle Program at MIT describing Toyota's business, coined the term lean in the late 1980s.

Lean Business Principles

The book *Lean Thinking: Banish Waste and Create Wealth in Your Corporation* introduced the American Business Market to lean business principled in the early 1990s. Lean thinking originated in the manufacturing models in Toyota automotive in the late 1980s following the introduction of Kanban. Different businesses and industries now use lean business models and principles to reduce time spent on delivering high-quality products and reducing the number of resources used to achieve that goal. Some principles of lean business models will now be detailed.

Perfection

It is important for a business to continually refine the first four principles to ensure that processes have minimal or no waste in them. The idea behind this principle is that any waste that goes unnoticed in the first four stages is always exposed over time. It is important to eliminate that waste to help a business adapt to the changing needs of its customers.

Peter Hines has argued that the five principles of lean thinking may be insufficient for some or most contemporary business situations. He stated that businesses need to apply lean thinking only to some processes like order fulfillment without giving any regard to communication, leadership, or quality management. Therefore, it is essential to understand how lean thinking can be applied to help a business develop a holistic approach to the delivery of products and services.

Value Identification

Every business and team must bear in mind that the value of any company will only lie with the customers or stakeholders. Every company should strive to meet the customers' demands. If the customer requires a specific product, the business must work on identifying the right resources to complete that job. Every business needs to understand the customers' needs and identify the products or services that will cater to those needs.

Value Stream Mapping

Once the business defines the products or services that it should provide to the customers, the management needs to map the different processes that the teams will need to complete to deliver that product or service. The business can identify the different steps included in the project during the mapping process. Businesses can also identify the steps that do not add any value to the project and eliminate those. For instance, if the business discovers the process to place orders by employees is complicated, it must eliminate that process since it is a waste contributor.

Flow

As mentioned earlier, the business can identify the various steps that do not add any value to the processes during the mapping step. The business should then work on removing those processes from the system. This will ensure that there are no obstacles in the business, and the product or service can be delivered to the customer with fewer hiccups. For instance, if a gardening service must visit an off-site location to stock up on supplies, it will take a longer time to deliver its services. The business should try to identify whether they can afford to increase the on-site storage space.

Pull

Lean processes always produce the output based on the demand of the customers, which makes these processes "pull processes". A pulled process is one where the call for the production of a product or service is on an as-needed or as-wanted basis. In service businesses, the delivery is always dependent on the workforce. For example, a pizza delivery service can choose to hire delivery executives based on the demand for pizza. If it is the football season, more customers are bound to order pizzas, and it is a good idea for the business to hire more executives to deliver the pizza.

Lean Tools

Lean companies use the support of different tools and processes to support the principles of lean. These tools help the business identify the processes that do not add value to the business and remove those processes. This chapter lists some of the tools that you can use to implement lean in your business.

Takt Time

Lean businesses should always look for different methods to optimize processes. This is the only way the business can ensure that they satisfy the demands of their stakeholders or customers. Takt time is the average rate at which a team or business produces the output based on the customer's requirement and in the stipulated time.

Takt = (Time available to produce a product or service) / (Demands made by the customer)

Continuous Flow

This tool ensures that the batch size is reduced to eliminate some constraints in the system. The business must identify a method by which it produces products or information that moves at a consistent pace from one step that adds value to the next with zero delays or waste in between.

Standardized Work

The business should create a document that will list the processes or methods that the organization uses to produce the goods and services to meet the calculated takt time. This document will enable the business to standardize the tasks and improve the value of the workstream.

Kanban and Pull Systems

As you read earlier, Kanban is a project management methodology that allows the project manager and the teams to schedule tasks. It also sheds some light on the different processes that every department must complete to obtain the result. This tool was developed by Taiichi Ohno to improve the manufacturing process of Toyota automotive in the late 1980s.

Cellular Manufacturing

Cellular marketing will aim to reduce the time it takes for the business to meet the demand of the customer or market, and allow the business to identify the processes that reduce the time taken to deliver the necessary output. Workstations and equipment are arranged to bring different teams to manufacture similar products.

The Why's

It is important for a business always to question the problem and identify the cause of any problem that may arise in the process. For instance, if there is a delay in the supply of the raw material, the business should ask the following questions:

- Why is there no backup supplier?

- Why are we sticking to this supplier only?

Level the Workload

Customer patterns always vary in the business, and the processes in the business are always consistent. This means that the processes in the business can only cater to handling specific customer patterns. Businesses need to identify the different patterns and plan the workflow based on these patterns.

Problem Solving

Most businesses or teams adopting the lean project management method adopt the PDCA cycle. This cycle is a graphical and logical representation of how individuals in the company identify problems and solve them. This will allow the company to view the process at a granular level.

- Plan: Establish the plan to achieve the final goal

- Do: Implement the plan

- Check: Collect, collate, and analyze the results

- Act: Implement reforms only if the business is unable to obtain the desired results

The business can now work on developing a system that will help the employees identify the problems and solve them by identifying the cause of the problem. The business can also work on implementing some countermeasures that will eliminate the problem.

Chapter Five: Start a Lean Project

Now that you understand what the Lean methodology is, now look at how you can implement lean project management at work. You can implement lean methodology using the phases below:

- Initiation

- Planning

- Execution

- Monitoring and Control

- Closing

Initiation

Before you begin a project, you should work on defining the value or the objective of the project during the project initiation phase. This project charter will specify the value of the project and also the items of value that the team should produce at the end of the project. This project charter will define the project requirements, including the quality standards, support requirements, interim deliverables, and maintenance required for project success.

Planning

It is during the planning phase that the project manager will define the value steam and map that value stream to different tasks. When it comes to lean project management, it is a good idea to use a work breakdown structure (WBS) to create the value stream map since this identifies the different steps that the team will take to complete the project. Lean project management will develop a schedule that will specify the timeline for the entire project.

Execution

The lean project management excels in execution. The project execution phase of lean management will include the following steps:

Flow

Once the tasks enter the production process, the goods being produced—like information products, tangible goods, services, etc.—do not stop moving. For a project that does not require full-time work, the project manager and team should work together to identify and document the set of partial deliverables at regular intervals. For instance:

- In an engineering project, the specifications and plans should be produced using the takt time as a criterion

- In a construction project, the tradespeople will line up and work on tasks without any gaps

- In a software development project, the functionality will be demonstrated using a continuous cycle, daily, monthly or whatever timeline suits the project better

- In a training project, the material must be produced at a measurement that works best for the team

Inventory

The lean methodology helps to minimize the inventory being used. This will ensure that the product delivered to the customer is better than what was expected. If there is a task that the team can complete without any additional inputs, the work can be completed and delivered to the customer. For example:

- In an engineering project, every specification and drawing will be certified as complete without having to wait for the other specification and drawing

- In a construction project, team members should complete the tasks and the project and also accept the ownership of that task at some point during the cycle

- In a software development project, the owner of the task is given the authority to take care of product functionality for the acceptance in the stages instead of the entire product

Kaizen

The lean process is managed continuously to identify waste or any other process or task that does not add value to the project. In this stage, the project manager should monitor the work breakdown structure to look at the scope.

Monitoring & Control

Every team member working as part of a lean team has the authority to push the red button or stop any process if they notice a problem with the quality. This task will stop the minute the issue has been identified, and a solution is found. Every individual project or completed task will be given this treatment. The team will not move to the next task until the completed task meets the specified quality standards.

Lean project management will use jidoka, a lean concept, where the quality of the product is checked at every stage of the process, and any issues in quality are identified and rectified immediately. It is also acceptable for the team to stop working on any other task until the completed output meets the quality standards. Therefore, every task in a project is defined as complete using human judgment.

Closing

Lean project management will seek to hand the deliverables over to the customer very quickly. The team will factor in dates when they can deliver partial deliverables to customers at an earlier date to understand whether the stakeholders will accept the product or not.

Chapter Six: Lean Six Sigma

Lean and Six Sigma are different approaches, but they can be combined because they have some similarities that allow them to work well together. The differences ensure that there are some analytical solutions and tools that will help to improve the process, service, or product. It is because of these similarities that both lean and six sigma analysis can be performed on the same process, service, or product.

Similarities of Lean and Six Sigma

The similarities of these methods:

• Both methods rely on the definition of the value, which is based on the experience of the customer. The customer is the boss.

• Both these methods use a process flow mapping approach, which will enable them to understand the process better. Even if the analysis is only based on a process, product, or service, there is another process that will be associated with creating and delivering the service or product.

- The methods rely on the data, which is used to determine the current performance and also for determining the impact of the processes on the future performance of the business. The data collected from a Lean Six Sigma project can be used in both the Six Sigma and Lean analysis. The reliance of these analysis methods on the data will help to ensure that the cause of the problem is defined.

- Both methods are applied to projects that include cross-functional teams. The size of the team and the duration of the project will depend solely on the scale and scope of the process, service, or product that is currently being analyzed for the improvement.

- These methods were earlier used only on the manufacturing operation, but it can be used for all functions in the business. These methods can also be used for internally and externally facing processes. They are also used in different industries, including consumer, industrial, non-profit, education, and government.

- Any improvement made to the processes using either approach will reduce variation and will also reduce waste. Removing any waste steps and activities will help to remove numerous sources of variation. When you remove variation, you can remove the wasted process steps and capacity.

There are, however, some differences in these approaches. These differences will not create a conflict, but they will provide multiple paths that can be used to arrive at the same destination. A lean six sigma project will always identify the defect defined by the customer and dictate the appropriate tools to cater to that defect. The final solution is obtained as a combination of both Six Sigma and lean improvements.

What is the Difference between Lean and Six Sigma?

Now that you know what the similarities are, look at some of the differences between these methods:

1. Both methods focus on different aspects of the problem. Six sigma focuses primarily on reducing variation, which means that the processes will always need to work towards meeting the target performance levels, while lean focuses primarily on waste.

2. Each of these methods implements different techniques. Lean uses visual techniques to identify the solution and analyze the progress. The solution is derived from the data collected during the analysis. Six Sigma uses statistical techniques to analyze and identify the right solution by using data visualization. This will lead to the assumption that it is easier to use lean over six sigma since the visual analysis is easier to understand. People are often intimidated by the Six Sigma analysis. The truth is that both these analyses are easy to understand and perform with the statistical tools that are present today.

3. Different types of documentation are present to obtain a solution. The lean solution uses a value stream map that will address the changes in the workflow. If you follow the lean methodology, you should maintain a document that will provide instructions on the many steps in the process. The six sigma solution focuses on changing the setup procedures. It also talks about how to monitor the process and respond to any variation in the process. The changes will also impact the instructions and also lead to some changes in the measurement systems or approaches.

Both these approaches are complementary in many ways, and it is easy to merge them into one methodology. Lean Six Sigma will avoid the issues of the earlier approaches.

Lean Six Sigma principles

Some principles of lean six sigma, which makes the methodology effective, will now be detailed. There are many other principles of this methodology, but the ones listed in this section have always led to successful programs.

Addressing a Real-World Problem

Lean six sigma is both a bottom-up and top-down approach or methodology. The latter is associated with the selection of the problem. Most teams work on real-world problems that impact different processes and customers. When it comes to lean processes or any process in general, the team members have to collectively work on reworking or revising the output based on customer feedback. This will make it hard for the team members to focus on the other parts of the project. This will lead to a sense of urgency when it comes to completing specific tasks. It is important to understand that this is real work.

There was a program called Quality Circle that was adopted by many companies in the year 1980. In this program, teams in an organization were allowed to choose their projects. This does sound great since it empowers the employees, but there was an issue with the projects selected by these teams. Most teams did not choose real-world problems. Some teams may choose trivial problems like putting up curtains in the lunchroom or repainting the office. Teams viewed this initiative as a fun initiative and did not worry about improving the business.

It is hard to make businesses and teams understand the importance of this method, which makes it difficult for most businesses to succeed in using this method. It is easy to convince teams to adopt this method if they understand why it is important for them to identify the problem and work on a solution to fix that problem. The management should not define both the problem and the solution. The teams should analyze the problem and determine the cause.

A Team Accomplishes Analysis

Lean six sigma teams often have members with different capabilities, and each of the members is given a different problem to analyze. Business processes are often cross-functional, and teams must perform cross-functional analysis to optimize and improve the process. It does not make sense to improve one step at the expense of other steps. This does not remove any variation or waste, and it just shifts the focus to a different step in the process.

Some issues arise with the lean six sigma approach since employees with the Green and Black belts often chose to identify the problem and also develop the solution to that problem without taking the inputs from the other team members. This would be an effective method if the teams were very small, and the project manager and other team members understood the entire process. This method, however, will not work for cross-functional teams and projects. If the project leader or manager is determined to identify the problem or solution without the help of other team members, and he or she does not know what the problem is, the project will be delayed.

When you include a cross-functional team, the perspective of every department in the organization is involved. These teams will collectively work together to understand the problem and develop the solution to the problem. Every member of the team must have a clear understanding of the problem. The problem must be looked at from different perspectives since that will help the team develop a solution that will address the issue immediately. This solution will also eliminate variation and waste in the other steps in the process.

The Analysis is Focused on a Process

Lean six sigma is an approach that one can use to analyze processes. This approach is more effective when it is applied to processes that either build or design products instead of looking only at the product, even if the problem is a product problem. Most businesses use the lean six sigma analysis of the "improve or investigate" processes and actions. Actions or tasks are a part of the process, and teams can complete most of these tasks without worrying about whether they are preceding or succeeding tasks. The teams should only focus on the context in which these tasks occur in the process. You can use the Six Sigma process map or lean value stream map to obtain a picture of that process.

Experts say that a process map helps teams immediately understand what is happening in the project. These maps also help teams recognize the problems in the process, which are often hidden since only an individual is aware of the task that he or she is performing. Your customers can come back with feedback about a defect in the product, which means that the team must rework on the product and fix that defect. Instead of focusing on the process that was followed to develop the product, the teams focus only on the defect. This is where they make a mistake since the team could identify the defect in the product, but they could not identify the cause of the problem. Teams can only do this when they define a process map and identify the problem.

The Analysis is Based on Data

Lean six sigma focuses only on data and not intuition. The value stream map will have a list of all the steps in the process, and the team collects data at every step. It is during the measure phase that the current condition of the product, service, and process is calculated. In this phase, the team should measure the defect or problem and measure whether the steps are being followed correctly. The data collected will help the team, and the project manager determines the actual state of the problem. This analysis will help teams identify the underlying causes of the problem and correct that problem. This is not where the teams stop relying on the data. Once the team identifies the solution, the team should collect data to assess whether the solution has fixed the problem or not. Once the data confirm that the solution works, the team should continue to collect data to ensure that the problem does not return.

One of the main challenges that most teams have with problem-solving initiatives and continuous improvement is that they are unable to accept the current conditions. Most businesses are in denial about many issues and problems in the company. Look at an example: There is a process in a company that is working on resolving any product issues that created large levels of rework for the team because there were numerous complaints from the customer. The business worked on solving the issue by fixing the problem with the product on many occasions by identifying the task that led to the defect. The team did identify the problem, but the issue here was that the team did not look at all the other steps in the process. If the team paid better attention to the data collected from the processes, the members would have identified that there was more than one task that led to the defect in the product. It is important to ensure that the data is fully understood and looked at correctly.

Understand the Impact of the Process Sigma

This principle focuses on the use of the six sigma analysis. Sigma is a statistical term that represents the variation that occurs from the normal values in the business. It is tied to the characteristics of the parameter that the business uses to measure success. There will be no variation between the same attributes of a process or product. Regardless of the number of times this process or product occurs, the attribute will not change. There will be some variations in other attributes. There will be an average value and some uncertainty assigned to the occurrence of some instances. Sigma represents uncertainty.

- One sigma is used to represent the boundaries of uncertainty for two-thirds of the occurrences

- Two sigmas are used to represent the confidence interval for 95 percent of the occurrences

- Three sigmas are used to represent the confidence interval for 99 percent of the instances

- When you reach the six sigma figure, you will note that there are only three occurrences out of a million that the normal variation can make a change to the product or output

It is important to understand that sigma only talks about variation and not acceptability. If you read the paragraph above, you will notice that it is not talking about assessing whether the customer accepted the product or not. The attribute in question can have a very small value for sigma, which means that there is no variation. That said, if the mean or average value of that attribute is out of the acceptable bounds for a customer, the product or output is defective. If the attribute has a high sigma value, it means that there is high uncertainty associated with that variable. If the customer does not have any expectations from that attribute, the product will still be accepted.

The Lean Six Sigma project management methodology does not use the sigma value to assess customer acceptance. It looks at how to accommodate for any high variations in the attributes. You need to understand that your output is generated through a process. When there are processes with high variation and high uncertainty, the business will need to spend more time and money to assess the process. You must remember that you are following a process, and the outputs of the previous step become the inputs of the next step. If there is too much variation in the input variable, it is important to develop the process in a way that the consecutive steps can accommodate different possibilities and values for those inputs. This will add some complexity to the processes. When you lower the sigma level, you can simplify the entire process.

The Solution Addresses the Real Root Cause(s)

Lean Six Sigma is one of the most common and powerful continuous improvement and problem-solving methodologies since it will identify the different characteristics of the actual problem. Some methodologies look at the problem from a single point of view. They will identify the problem, look for the cause, and develop a solution to address that problem. These methodologies work on the assumption that the solution will eliminate the entire problem. Other methodologies work on the assumption that the problem occurs in almost every process, and it occurs very often. The process is inadequate or flawed, and a change to the process will eliminate the problem.

The goals of both these types of methodologies are similar and quite admirable. The way to fix the issue in the first method is to place a correction in the process to control the cause. In the second method, the entire process will be reengineered. This does not improve the situation and often makes the situation worse. Lean Six Sigma uses different tools to identify the problem and assess if the problem is a common cause or if it is a special case. Once this differentiation is made, the team can work on identifying the cause of the problem and address that problem appropriately. The team can develop a solution strategy to address the problem. The team can implement a simple solution if the problem is a special cause. If the problem is common, the team can work on redesigning the process.

Benefits of Lean Six Sigma

Lean Six Sigma, like lean, is a continuous improvement approach or methodology. It is, however, important to understand how this method improves processes. Does this method increase profits or sales? Does it reduce the number of complaints from customers, or does it improve customer satisfaction? Does it lower the costs, improve the quality of the raw material or the product, or lower the cost of quality? Does the process improve employee morale? Can you use this process to promote your products and services? Does it increase benefits and pay? Well, yes. Now, look at some benefits of the Lean Six Sigma methodology for a business and an individual.

Organizational Benefits

Since lean six sigma is a continuous improvement framework or methodology, an organization is bound to benefit from this method. General Electric (GE) claims that the lean six sigma methodology has helped the company save over $2 billion. Look at some of the benefits of this method and its implications.

Simple Processes

Lean six sigma will help businesses and teams simplify numerous processes. Since the methodology focuses on a cross-functional approach, the value stream maps will help the organization identify the inefficiencies and waste in different parts of the process. Many processes will embed workarounds and rework to cater to persistent problems. If the team can identify these workarounds or areas where there is a lot of rework, they can remove the waste, which will then make the process easier to control and manage. This will lead to the creation of a faster process, which will then lead to higher customer satisfaction and better customer service, thereby increasing sales. Additionally, faster processes will reduce any overhead costs, thereby increasing profits. Simpler processes have fewer errors, and it is for this reason that fewer defects and higher quality characterize these processes.

Fewer Errors and Mistakes

Now dig a little deeper into this benefit. The Lean Six Sigma methodology defines the quality of the product based on what the customer values. When businesses focus on improving processes to meet the external requirements, they will address those problems that have an impact on the success of the business. The teams must use the data collected at every step and use that data to improve some real problems in the organization. So, the Lean Six Sigma method does not just cater to improving processes with mistakes or errors but also focuses on improving the processes that matter most to the business.

Predictable Performance

It is easy to manage and control simple processes when compared to complex processes. In addition to the above benefits, Lean Six Sigma also focuses on removing any variation in the processes, thereby making the processes predictable. This means that the teams can predict the cycle time, costs, and quality of the output. This level of predictability will lead to higher profits, better customer service, and fewer complaints. Organizations that have this level of predictability can work well in an environment where the changes are fast-moving. Changes in customer expectations and technology will create an unstable business environment. If you do not have predictable processes in the business, it will be impossible to develop a solution to cater to this instability.

Active Control

The Lean Six Sigma approach can help businesses and teams actively control the processes. The methodology reduces the cycle times and uses real-time data to analyze these cycles allowing the businesses to develop real-time data systems and control plans. Process managers and operators can make the right decisions that directly impact the performance of the processes. This will improve employee morale, agility, and performance. An operator should understand how their work improves or impacts the performance of the process. The lean six sigma approach gives the operator immediate feedback. Since the operators working on the task are given control of these processes, they do not feel like the victims. With active control and short cycles, an organization can respond to the changing markets and grab new opportunities.

Personal Benefits

You, as an individual, can reap the benefits of lean six sigma within an organization. Now, look at some of the benefits of lean six sigma that you, as an individual, can expect when you work in an organization that participates in lean six sigma.

Personal Effectiveness

The lean six sigma project management methodology is a problem-solving methodology that can be used to address problems of different kinds. Your ability to perform in any industry or position in the business will improve if you learn how to identify and fix some problems. The lean six sigma project management method will steer you through an organized process of analysis, problem identification, inquiry, and solution creation. You can apply numerous tools and techniques to some of the common issues and problems. Even if you do not want to use all the tools that are available to you, you can use the lean six sigma approach to control the processes and identify the problem. You can use this method to solve issues in different business settings.

Leadership Opportunity

The lean six sigma project management methodology is implemented through numerous projects, and each of these projects has a different leader. When you lead a lean six sigma project, you will have the opportunity to look at different functions and speak to the senior management. This exposure will help you develop the abilities to identify a problem and also come up with a solution to fix that problem. When you interact with managers and team members, you can improve your decision-making skills and communication. The structure of the lean six sigma model will help you develop and work on your project management skills. You can seek the next opportunity or a promotion if you can put it on your résumé that you led a project that improved quality, saved costs, and also reduced the time taken to complete the project.

Industries and Functions using Lean Six Sigma

The lean project management methodology was first used in the engineering department by an automotive manufacturer. The Six Sigma project management methodology was first used in the quality department of a technology system manufacturer. These methodologies, however, have moved beyond these industries and departments. The lean six sigma project management methodology can be used in different departments, including:

- C-Suite
- Call Center
- Customer Service

- Design Engineering
- Field Sales
- Finance
- Human Resources
- IT
- Legal
- Logistics
- Maintenance
- Manufacturing Engineering
- Manufacturing Operations
- Marketing
- Process Engineering
- Purchasing/Sourcing
- Quality
- R&D
- Sales
- Test

The lean project management methodology has moved beyond the realm of manufacturing, and many businesses in different industries have begun to implement lean six sigma in their business processes. In some cases, the business may focus more on six sigma or lean, but many businesses focus on a combination of lean and six sigma.

- Agri-business
- Aviation
- Banking
- Electronics

- Financial Services
- Government
- Higher Education
- Hospitals
- Manufacturing
- Medical Devices
- Mining
- Oil and Gas
- Pharmaceuticals
- Retail
- Telecom
- Transportation

Chapter Seven: Lean Startup

Eric Ries said that startups could be a success if they follow a certain process. This means that the process can always be learned, and those who have experience can also teach them. Every entrepreneur will always wonder whether a startup will fail.

If you wish to begin a lean startup, you must identify a small gap in the market using time and money effectively. You will need to use different techniques to ensure that your product or service reaches the market in a faster way while also avoiding the production or manufacture of products that no consumer will want.

Most amateur entrepreneurs feel that they are taking a shot in the dark when they are identifying a product or service they can offer to their potential consumers. However, it does not always have to be a trial and error proposition. If you adopt lean thinking, you will be able to develop ideas and refine them to meet market standards.

The following are some principles that will give a startup a greater chance of making a profit and becoming a success within a limited budget.

Principles of Lean Startup

Validated Learning

A startup does not exist only to build products for the customers or to make money. It also can only exist when the management learns how to build a sustainable business. The learning can be validated through statistic measures by running experiments that test the startup's vision.

Entrepreneurs are Everywhere

Eric Ries believes that every individual in the world is an entrepreneur. Some successful entrepreneurs have built their organization in their garage. You can find entrepreneurs in Hollywood, in the IRS, and even in well-established organizations. These people are always looking for a way to develop products that increase value to the customer.

Controlled Use and Deployment of Resources

One of the most important principles of a lean startup is that the startup must use every one of its resources effectively and efficiently. Since most startups do not have enough investment, they use the lean business model to encourage the effective deployment and continuous development of the resources that the company does have.

A lean startup must continuously evaluate how the initial investment can be used to meet its targets and customer requirements. The startup must also ensure that it does not spend more than what is necessary to test, evaluate, and refine its products. If the costs are kept at a minimum, the startup can maximize its profits whenever there is a sale.

Every lean startup is dependent on organic growth since it does not have a huge capital investment. When the profits made at the early stages are reinvested in the company, the startup can scale its operations up in a controlled manner without sacrificing quality. This is commonly called innovation accounting.

Innovation Accounting

A startup must focus on the following to improve outcomes and also hold every entrepreneur accountable:

- How can progress be measured

- How can milestones be set

- How can work be prioritized

Entrepreneurship is Management

It is important to remember that every startup is not defined by its products but is an institution. Therefore, there must be a management team in place to understand and develop the startup.

Build-Measure-Learn

Every startup looks for ways to convert its ideas into a product or service and measure how its customers receive that product or service. When they understand the response, they will understand whether they need to pivot or persevere. This process is covered in further detail in the next chapter.

Models and Methodologies

The lean startup model was introduced in 2011, and its impact on the economy has been enormous. The book written by Eric Ries gained immense publicity, and many companies use the information in the book to develop their startups. However, the ideas in the book are not new; most entrepreneurs have forgotten these ideas since success is always measured in numbers in the business world. The methods and ideas in the book are valuable to startups as well as well-established organizations. In his book, Eric Ries has defined a startup as a human institution whose goal is to create a new service or product under uncertain conditions. This chapter covers some of the common methods used by lean startups to design products and services of great value for the customer.

Build-Measure-Learn

The way different companies pursue innovation in today's market has been affected by the idea of using certain scientific or statistical methods to handle or calculate uncertainty. This means that the company must define a hypothesis, build a product or service to test that hypothesis, use that product or service and learn what happens, and finally adjust the attributes of the product or service to increase the value for customers.

The Build-Measure-Learn methodology can be applied to almost anything. You do not have to use this methodology to test new products alone. You can also test a management review process, customer service idea, new features to existing products or website offers, and tests. You have to carry out a test and validate the initial hypothesis to ensure that you have enough data to assess the value of the product to the customer.

The aim of every company should be to move through this methodology quickly. You have to identify if the product or service developed is worth going through another cycle or if you should come up with a new idea. This means that you must define a specific idea that you want to test with minimum criteria that can be used for measurement. When it comes to products, you have to test whether your customers want to purchase your product or if they need it. You have to learn what your customers want and not trust what they think or say they want.

Minimal Viable Product (MVP)

A traditional company will first have to define the specifications of every product it wants to produce or manufacture and then assess the high cost and time that will be invested to produce that product. The lean startup methodology encourages every startup to build the required amount of product through one loop of the Build-Measure-Learn loop. If the company can identify such a product, it becomes a minimal viable product. This product is manufactured or developed using minimal effort and less development time.

Every startup does not necessarily have to write a code to automate processes to create an MVP. An MVP could be as simple as a slide deck or design mockups. You have to ensure that you run these products by your customers to get enough validation to pass this product through the next cycle.

Validated Learning

Every startup must test or validate a hypothesis with the right idea in mind—learn from what is observed. There are times when startups have focused on vanity metrics that made them believe that they were indeed making progress. This is not the right approach since you must always look at metrics that will give you some insight on the product and how it can be changed to increase its value to the customers. For example, the number of accounts opened on Instagram is a vanity metric for that platform. The actual metric would be the number of hours spent scrolling through Instagram by each account holder. This will give the developers the true value of the product.

In the book Lean Startup, Eric Ries has provided an example of his own. A company called IMVU always showed a chart that painted a good picture of its management and investors. Many registrations were being made every single day. However, this graph did not show if the customers or users value the service. The graph did not show the costs that went into marketing to acquire new users. This chart only looked at vanity metrics and was not designed to test a hypothesis.

Innovation Accounting

Through innovation accounting, a startup can prove that it is learning to grow and sustain itself as a business. A company can do this in the following ways.

Establish a Baseline

The startup can run an MVP test and collect data that will enable it to set some benchmark points. You can use a smoke test where you can market the product or service you want to offer and assess your customers' interests. This includes a sign-up form to understand if the customers want to purchase the product or service. Using that information, you can set the baseline for the first iteration of the Build-Measure-Learn cycle. It is all right to make mistakes or have low numbers since that will help you improve.

Tuning the Product

Once the baseline has been established, you should identify one change that must be made to the product and test that improvement. Do not make all the changes at once, as it can lead to chaos. You can try to see how a change in the design of the form attracts more customers when compared to the earlier design. This step must be carried out slowly to ensure that every hypothesis is tested out carefully.

Persevere or Pivot

When you have made several iterations through the cycle, you have to move up from the initial baseline towards the goal that was set out in the business plan. If you are unable to reach that goal, you must learn why using the data collected at every step.

Pivot

A successful entrepreneur is one who has the foresight, the tools, and the ability to identify which parts of the business plan are indeed working for the company. They also learn to adapt to changes in the market and their strategies according to the data collected during the iterations.

One of the hardest aspects of the lean startup method is to decide to pivot since every entrepreneur and founder is emotionally attached to the product they have created. They spend a lot of money and energy to get to where they are. If a team uses vanity metrics to test its products and hypothesis, it can go down the wrong path. If the hypothesis selected is not defined clearly, then the company may fail since the management does not know that the endeavor is not working. If you, as the management, decide to launch the product fully in the market and then assess the outcome, you will see what happens, and there is a higher probability that you may fail.

If you choose to pivot, it does not mean that you have failed. It means that you will change the hypothesis that you started with. The following variations are often used when a startup chooses to pivot.

- Zoom-in Pivot: A single feature in the product that sets it apart from other products becomes the actual product.

- Zoom-out pivot: This is the opposite of the above definition, where an entire product is used as a new feature in a larger product.

- Customer segment pivot: The product designed was correct. However, the customers that were selected were wrong for the product. The startup can change the customer segment and sell the same product.

- Customer need pivot: When the startup follows the principles of validated learning, it will identify the problem that needs to be solved for the customer who was initially selected.

- Platform pivot: Most platforms start as applications. When the platform becomes a success, it transforms into a platform ecosystem.

- Business Architecture Pivot: Based on Geoffrey Moore's idea, the startup can choose to switch to low margin and high-volume products from the high margin and low-value products.

- Value Capture Pivot: When you decide to measure the value differently, you will be able to change everything about the business right from the cost structure to the final product.

- The engine of Growth Pivot: According to Ries, most startups follow a paid, viral, or sticky growth model. It would be more prudent for the company to switch from one model to the other to grow faster.

- Channel Pivot: In today's world, advertising channels and complex sales have reduced since the Internet provides a huge platform for a company to advertise its products.

- Technology Pivot: Technological advancements are being made every day, and any new technology can help to reduce the cost and increase performance and efficiency.

Small Batches

There is a story where a father had asked his daughters to help him stuff newsletters into a document. The children suggested that they fold every newsletter, put a stamp on the envelope, and write the address on the envelope. They wanted to do every task one step at a time. The father wanted to do it differently—he suggested that they finish every envelope before they moved on to the next envelope. The father and daughters competed with each other to see which the better method was.

The father's method won since he used an approach called "single-piece flow", which is common in lean manufacturing. It is better to repeat a task over and over again to ensure that you master that task. You will also learn to do the task faster and better. You have to remember that an individual's performance is not as important as the performance of the system. You lose time when you should go back to the first task and restack the envelopes. If you consider the process as a unit, you can improve your efficiency.

Another benefit of using small batches is that you will be able to spot the error immediately. If you fold all the newsletters and then find out that that newsletter does not fit into the envelope, you will need to fold all the newsletters again. This approach will help you identify the error at the beginning and improve your process.

The advantage of working with small batches is that you will be able to identify the problems soon.

Andon Cord

The Andon Cord is a method that was used by Taiichi Ohno in Toyota, which allowed an employee to stop the process if he or she identified a defect in the process. If the defect continues longer in the process, it is harder to remove that defect, and there is a higher cost involved. It is highly efficient to spot the defect at an early stage, even if it means that the process will need to stop to address the defect. This method has helped Toyota maintain high quality.

Eric Ries mentioned in his book that the company IMVU used a set of checks that ran every day to check if the site worked accurately. This meant that they were able to identify and address any production error quickly and automatically. There were no changes made to the production until the defect was addressed. This was an automated Andon Cord.

Continuous Deployment

Continuous deployment is a scenario that is unimaginable and scary for most startups. The idea of this method is that the startup must update the production systems regularly.

IMVU was regularly updating its production system by running close to fifty updates. This was made possible since they invested in test scripts. Over 14,000 test scripts would run at least 60 times a day and simulate everything from a click on the browser to running the code in the database.

Eric Ries also talks about Wealthfront, which is a company that operates in an environment regulated by the SEC. However, this company practices continuous deployment and has more than ten production releases a day.

Kanban

Kanban is a technique that was borrowed from the world of lean manufacturing. It was developed by Taiichi Ohno in the late 1980s to improve the manufacturing unit of Toyota automobiles. Eric Ries mentions the company Grockit, which is an online tool that helps one build skills for standardized tests. This tool creates a story in the development process, which is then used to develop a feature. They also mention to the user what the outcome or benefit of the tool is. These stories are validated to see how they work for different users. A test is conducted to see how this tool benefits the customer. There are four states:

- Backlog: The tasks that can be worked on but have not yet been started

- In Progress: The tasks that are currently being developed

- Built: The tasks that have been completed and are ready for the customer

- Validated: Products that have been released and have been validated by the customers

If the story fails the validation test, then it will be removed from the product and produced again. A good practice would be to ensure that none of the buckets mentioned above have more than two projects at a given time. If there is a project or task that is in the built bucket, it cannot move to the validated stage until there is enough room for it. The same goes for the processes that are in the backlog bucket. These tasks cannot move to the "In Progress" bucket until it is free.

A valuable outcome of this method is that the team can start measuring its productivity based on the validated learning and feedback from the customer. The team will then stop measuring its productivity based on the number of new features developed.

The Five Whys

Every technical defect or issue has a human cause at its root. The five whys technique will allow the startup to get closer to the root cause. This is a deceptively simple technique but is powerful. Eric Ries has mentioned in his book that most problems or issues that are identified in a process are caused due to a lack of training. These problems may look like an individual's mistake or a technical issue. Ries uses IMVU as an example to explain this technique.

- A new product feature or release was disabled for customers. Why? The feature tanked because of a failed server.

- Why did this server fail? There was a subsystem that was used incorrectly.

- Why was that server used incorrectly? The engineer using the server was not trained to use it properly.

- Why did he not know how to use the server? He was never trained.

- Why was he not trained? His manager did not believe that new engineers needed to be trained since he believed that he and his team were too busy.

This technique is extremely useful for startups since it helps them make improvements within a short period. A huge amount can be invested in training, but this may not be the optimal thing to do when the startup is still at its development stage. If the startup takes a look at the root cause of every problem, it can identify the core areas that need to be worked on and not focus only on the issues at the surface.

Most people tend to overreact to issues that happen at the moment, and the Five Whys help them understand what they need to look at to understand what is happening. There is a possibility that the Five Whys can be used as a way to blame people in the team to see whose fault it was. The goal of this method is to identify problems that are caused not by bad people but by bad processes. Every member of the team must be in the same room when this analysis is made. When blame needs to be taken, the management must take the hit for not having a solution at the system-level. Good practices to follow to get started with this methodology are:

- Mutual empowerment and trust. If a mistake is made for the first time, you should be tolerant of them. Ensure that you do not make the same mistake twice.

- Maintain focus on the system since most mistakes happen due to a flaw in the system and ensure that people always solve problems at the system level.

- The company should always face some unpleasant truths. This method will bring out some unpleasant truths to the surface, and the management should ensure that these issues are taken care of at the initial stage. If this method is not conducted in the right manner, it will change into the Five Blames.

- Always start small and be specific. You have to look at the process in detail and always start with small issues. When you understand the issues, you must identify the solutions. Always run the process regularly and involve as many people from the team as you can.

- Appoint someone who is a master at Five Whys. This person must be the primary change agent and should have a good degree of authority to ensure that things get done. This person will be accountable for judging whether the costs were made to prevent or work on those problems that are paying off or not.

The Five Whys methodology is used to transform the startup into a more organized and adaptive organization, which can be hard.

Chapter Eight: Lean Enterprise

You know what lean is, and you have learned about what a lean startup is. So, what do you think a lean enterprise is? A lean enterprise is an organization that is looking for ways to continuously improve processes. So, it is clear to understand what lean enterprise management is. Lean enterprise management ensures that the organization adheres to the principles of a lean enterprise. In simple words, the manufacturing process should indeed be lean, but the concept of lean should spread across every process in the organization. For instance, the commercial department should always ask itself if there is a better and faster way to give the customer the correct response the first time itself. The purchasing department should ask itself if there is any way they can ensure that the products they purchase are good the first time. Manufacturing is the core of every organization, but the speed of the process is related to every other function in the business. It is for this reason that you can develop a lean enterprise only when you spread the concept of lean across the entire organization.

How to Start to Create a Lean Enterprise

If you want to understand the concept of a lean enterprise better, watch the following video: https://www.whatislean.org/. The concept of a lean enterprise has been explained in detail. Now that you are aware of what a lean enterprise is, you know that all the departments involved in a lean company should also be lean. You may now wonder how human resources or finance departments can be lean. Yes, these departments and every other department in the organization can be lean. They will, however, need to have their own goals, metrics, and board.

The departments should also ensure that they develop metrics that are relevant not only to the customer but also to the company. So, now look at how you can create a lean enterprise. The easiest way to do this is to place a board and write the goal for the department at the center of that board. The only thing you will need to do is follow the steps of lean management and ensure that the department is now lean.

The Difficulties of the Flow of Information

The only reason every organization starts changing the functioning of the manufacturing department is that you can always see what you are trying to improve, and you can see an immediate effect. You will need to be careful when you begin to work with intangible information like:

- Phone

- Mail

- Call

- Chats

- Documents

- Fax, etc.

It is harder to work with intangible information because this information is from multiple channels and at a great speed. You can deal with this intangible information in the following manner:

- Create the standard or the process for the flow of information.

- Draw the process map and include the people involved at every stage of the process.

- Ask the members of the team whether the information is necessary.

You should at least start with the following step: Ask the team who will consume the information that is being collected. You should create the standard or the rules that will define who the supplier of the information is, who the customer is, and what the customer may want or need. You can use the SIPOC tool to perform these steps.

SIPOC to Standardize the Information Flow

Now look at how you can build SIPOC to manage the information flow in your enterprise. Seven steps will help you manage the flow of information:

1. You should first identify the process that you wish to map and define the boundary and the scope of that process. You should ensure that you use the right words to describe this process and also assess the time you will take to complete the tasks under each process. Make sure that you list the start and endpoints.

2. Make a note of the outputs. You should ensure that you list the products and the services that you will need to produce at the end of the project.

3. Ensure that you name the customers, end-users, or the stakeholders by their name and system. You should ensure that you map the correct output to the stakeholder.

4. Now, work on determining and defining the requirements.

5. Understand what the customers expect from you and what they demand from you. Make sure that you define these based on the budget that you receive.

6. Define the inputs you will need for the process. Identify the raw material, the information, capital, human resources,

and natural resources that you will need to use to produce the outputs.

7. Identify the sources of the inputs.

If you want to ensure that you follow a lean process, you will not only have to develop a lean manufacturing environment but will also need to develop a lean enterprise. You should then think broader and work towards changing the point of view of the departments. You can do this by placing a metric board in their offices. This may sound strange, and the people may push back, but you should make an effort. Once you see that there is a change in the way the information flows, you can use SIPOC as the main tool.

Chapter Nine: Lean Teams

The lean project management methodology has changed the way businesses work. This method will help to improve processes for the better. A lean team will enable a company to facilitate a positive change in the way the business manages the process, and will respond to any problems promptly. A lean team is a group of people who can make quick decisions and also take actions that will benefit the company.

Developing a Lean Team

It is important to create an effective lean team. To do this, a business should define the teams around the processes that the teams perform. For these teams to be effective, they will need to include people with diverse backgrounds and capabilities. These people should know how to improve processes.

Forming a Team

Once the project manager or the business has defined the various processes in the business, it is time to develop the team and focus on improving every process. The team should have a set of individuals with diverse capabilities and preferably not from the same department. Look at the example of engineering, procurement, and construction company. This company's current process results in some overpriced bids, which will mean that the company loses the bid. Alternatively, the company could have underpriced bids, which will reduce the revenue of the company.

The current issue can be because of the result of poor communication between the business and the stakeholders during the proposal process. A lean team can improve this communication and also lead to some positive change by including people from the different departments involved in this process. Instead of making any changes based on one team's perspective, if the changes are made based on the input given by different individuals, the changes made to the process will be made based on the input from different departments like:

- The field service team

- Project management

- Purchasing

- Engineering

- Logistics

When you form a lean team with members from different departments working on a specific process, the team can identify the problems and also develop a targeted solution to improve that process.

Empowering the Team

A traditional business structure will result in a vertical power structure where minimal input is received from the employees, which makes it difficult to improve the processes promptly. Lean teams work on addressing this problem by empowering the teams to make the right decisions to facilitate the change. The roles of every team and the responsibilities of every member will be clearly defined. These members can also work on improving processes. The teams must have access to make changes without changing the authority structure.

Lean Team Hierarchy

A lean team also follows a hierarchy similar to the organization hierarchy. This hierarchy will include all the levels of business. There will be a group leader who will facilitate the improvements and communication between the different teams. Each team will have a team leader who is responsible for implementing the changes to the processes and for obtaining feedback from the group leaders. Every team member is responsible for identifying a problem and looking for a solution to that problem. When a potential solution is identified, the team members will work together with the team leader to develop the solution.

One key aspect of a lean team's structure is that every group is empowered to implement some improvements within the scope of the project. If a single improvement impacts multiple groups, the group leader will bring the idea to the upper management for any additional support or approval. This structure will help to ensure that the team can implement a change to the process.

Implementing a Lean Team

If a project manager wants to implement a lean team, they will need to speak to the head of the organization since the implementation will need a top to bottom approach. Every manager, supervisor, employee, or worker must be involved in making the change. A worker or employee should expect the practices to improve, and the management should always support the change. To do this, there must be a fundamental shift in culture. The manager should be trained to lead and facilitate the change, and the team members must learn to improve these processes. It will become slightly difficult to make changes to the workplace culture. It is a good idea to use tools like the Five Whys, Kaizen, or the PDCA (Plan-Do-Check-Act) for this.

Kaizen

Kaizen is a continuous improvement philosophy. If a business wants to use this philosophy, it must ensure that every worker in the firm, right from the CEO to the assistants, is involved in the process. Kaizen will help to change the culture at a workplace by allowing workers to improve processes regularly and allowing the management to support that change.

Five Whys

The five whys will enable the worker to identify the cause of the problem and then fix it. This technique will ensure that they do not focus only on the superficial issues since that will not solve the problems. As the name suggests, the employees should ask "why" until they identify the cause of the problem.

PDCA

PDCA is a lean tool that helps to resolve any issue through four steps: Plan, Do, Check, and Act. When a problem is found, this tool will enable the team member to address it by identifying a solution, testing it, obtaining feedback, and applying it.

How to Build a Lean Team

Eric Ries, in his book *The Startup Way: How Modern Companies Use Entrepreneurial Management to Transform Culture and Drive Long-Term Growth*, describes the mantra of a lean team as "Think big, start small and scale fast." Regardless of whether you are working in a startup or working on building an internal project in an established organization, the lean team approach will maximize the efficiency and the results in uncertain times. That said, it is hard to assemble a group that can execute this vision. This process also comes with the routine challenges and questions that are unique to the methodology. Now, look at seven steps you can use to build a Lean team.

Start Small

Amazon follows the "two-pizza team" approach. In this approach, you must always start with a small team if you want to work on developing new methods. You should aim to develop a team that you can feed easily with just two pizzas. When you have a smaller team, you will see that the members bond faster, which will improve the communication within the team. A small team also ensures that a decision is made quickly, and new methods can be tested faster. There is also better accountability since every member of the team is aware of what they need to do.

Make the Team Cross-Functional

Yes, there are very few people on the team. This does not mean that you do not capitalize on their abilities. Every lean team should be cross-functional, which means that different members of the team should bring out a different ability or skill that will represent the different departments in the company. In enterprise organizations, the teams have employees from the same departments, and once they complete their work, the results or the output will be shared with the next department. This is an inefficient approach since ideas are not shared between departments, which will lead to subpar solutions.

If you want to build a cross-functional team, you should first sit down and understand the needs of the project. You must understand the project and identify the different departments that must be involved to make some progress. You should also identify potential roadblocks and see how they can be avoided. Eric Ries, in his above book, talks about an industrial project. For this project, the team should include a product designer, a member with manufacturing expertise, and marketing or a salesperson who understands the needs of the customers. A project in a different industry will require a different set of people. There are numerous combinations that you can look at, depending on what needs to be achieved at the end of the project.

Every project manager or team leader must be aware of what the project is and also see if they need to obtain some permissions from the legal department. Make sure that you identify the different departments that need to be involved in the project at the start so you can avoid any delays. You can always ask for volunteers if you have issues with finding a person from a specific department to join your team.

Never Over-Rely on Team Players

Most project managers make the mistake of depending on the same employees to ensure that the team works together. This will impact the productivity and satisfaction of those employees since they will be overloaded. A study was conducted by the Harvard Business Review to understand employee satisfaction. The study concluded that employees who are always in high demand because they are seen as collaborators in their company have the lowest career and engagement satisfaction scores. Some experts say that it is easy to prevent overload by reducing some unnecessary meetings, and let individuals know that it is okay for them to say no and let someone else take their place.

Train People to Be Team Smart

It is important to ensure that every employee in a team excels in that team. To do this, you must invest in training. Companies make the mistake of focusing on helping a team member develop professionally at an individual level. They develop training programs that do not focus on teams, but only on people. Managers and employees are never educated on how they can contribute effectively to the team or how to build a better team. Many companies are team dumb since the collective intelligence of the team is independent of the intelligence of the members of the team.

A company with the smartest employees can still have terrible teams. A paper published in 2010 in the *Science* journal showed that the collective intelligence or "c-factor" is correlated with the communication and environment within the team. This factor is dependent on how the conversations take place in the team, the social sensitivity of the group, and the number of women in the group. This research also suggested that teams that fail at completing one task are likely to fail all other tasks as well. You, as the project manager, can increase the c-factor in the team by guiding the different members of the team about how to work together.

Creating a Pro-Risk Environment

If you want to create breakthroughs or find some innovative solutions, you need bold ideas and the willingness to make mistakes. Individuals in lean teams should learn to welcome both failure and risk. It is difficult to create this mindset in teams since most organizations still follow the principle of "failure is not an option". The dynamics of the team will make it hard to change this mindset since every member of the team will want to play it safe. Nobody will ever want to look like a fool in front of their colleagues. That being said, you could use some pragmatic tools and psychological insights to coach your team into feeling brave about failing and taking risks.

Each member of the team has a different trait, but since everyone is a human being, most of their behavioral and psychological patterns are the same. This may seem obvious to you now, but the truth is that people often overlook this insight. Companies only focus on the personalities, capabilities, and expertise of the individuals they hire, but research states that people with different capabilities can work together and deliver projects on time if the right environment is created for them.

It is important to let your teams know that their decisions will not result in litigation or a loss of millions of dollars. You must help your teams understand what can be undone and what cannot. A team should have a reverse button at some point where they can step back, accept that it is not going to work, and try a different approach. They should, however, make those decisions quickly. If the team works on identifying the reversible risks, people will not be bogged down because this will reduce the chance of a blame game.

Understanding the Needs of the Team

Every member of a team will join that team with specific assumptions in mind, and they work on trying to understand how to get their work done. They also have some assumptions about how the communication in the team should work. If you want to ensure that your team works as a cohesive unit, you must understand the different assumptions that each individual is working with. Every team member walks around the team with an assumption about how each member of the team should behave. If there is an individual who constantly interrupts people, the other members of the team may believe that he or she is a jerk.

It is for this reason that experts recommend that every team should develop a charter of norms. These norms will answer simple questions like:

- How do we want to work together?

- How do we react to a situation where we disagree with each other?

- Are we going to make some proposals?

- Are there going to be arguments about tasks?

- How do we come to a decision?

When you develop the charter, identify the different scales that you want to cover. Ensure that you cover the scales surrounding evaluating, scheduling, communicating, disagreeing, trusting, persuading, deciding, and leading.

Measure to Learn and Improve the Team

Regardless of whether you want to build a new team or improve an existing team, you cannot simply start without considering the team and measuring the team. Your measurements do not have to be elaborate, and they can be as simple as assessing the current sentiment in the team. By this, you should try to understand how people feel about the team spirit. You can assess this during every team meeting. Ask your team members to give the team a rating between one and five. If they are not comfortable about sharing the rating in front of the entire team, you can ask them to rate the team on a piece of paper. When you receive the information, you must act on that information. For instance, if most members of the team have rated the spirit as one, you should spend some time to understand why they feel this way and work towards developing a solution to cater to the problem. When you do this, you apply the build-measure-learn lean startup methodology to your team.

Teams do vary across a company, but a lean team will only be effective if it is small, and the members have diverse capabilities. It is important for you, as the project manager, to create ground rules, ensure that every member contributes, and check in with the team to assess how individuals feel about the team and the environment. From here, you can work on the feedback you receive and improve the processes. You must remember that you and your team should work towards continuous improvement, and this means that you can never accept that your team is perfect.

Chapter Ten: What Is Lean Analytics?

Lean Analytics is an innovative and great method to help you streamline the sales funnel. When you combine this method with the psychological insight of American author Nir Eyal's, you can guarantee success. From the previous chapter, you will have understood what the Lean Startup approach is and the different techniques and tools you can use to follow this approach. Lean Analytics is only an extension of the lean startup method. Alistair Croll and Benjamin Yoskovitz, in their book *Lean Analytics: Use Data to Build a Better Startup Faster*, wrote that a lean startup would help you structure the progress and also help you identify the risky parts of the business. The lean startup method will also help you learn more about how you can adapt and overcome these risks. Lean Analytics is used to measure the team's progress and also help the business answer some questions to get the required answers.

What Kind of Business Are You?

Lean Analytics is one of the best ways to streamline sales funnel. The idea behind this approach is that when you know the type of business you are in and also the stage your business is at, you can track the important metrics and optimize those metrics. These metrics will be the ones that matter most to your business.

Croll and Yoskovitz described the five stages that every startup will go through if it wants to implement lean analytics:

Empathy

In this stage, the business should work towards understanding the customers better. The business should learn more about the problems that they want to solve and work towards building a project that will solve these problems.

Stickiness

When it comes to stickiness, the business should look at how it can develop a product to engage the customer.

Virality

In this stage, the business should focus on user acquisition. In simple words, the business should try to identify the onboarding process.

Revenue

The business will now need to focus on monetizing the product.

Scale

It is finally time for you to expand your sales and diversify.

How to Apply Lean Analytics

Lean analytics, like every other approach, has its canvas. This is an across-the-board tool, and you can use this tool for different methods or processes. A Scrum expert, Nicolas Nemmi, developed the lean roadmap canvas based on the hooked model developed by Nir Eyal.

The Lean Roadmap

	Hook Model			
Lean Analytics Stage	Trigger	Action	Variable Reward	Investment
Empathy				
Stickness				
Virality				
Revenue				
Scale				

There are four boxes in the canvas developed by Nemmi, and each of these boxes is combined with the five lean analytics stages described above.

Trigger

The trigger will define what it will take to ensure that the user gets to the product.

Action

This will help you define the simplest behavior in anticipation of the reward or feedback. In other words, you are looking for a way to make it easier to remove any friction in the buying process. An example of this is Amazon's 1-Click ordering. This method accomplishes the goal.

Variable Reward

This will help you define if the reward is fulfilling, but leaves the user wanting some more. Candy manufacturers are brilliant at doing this.

Investment

What is the customer doing to ensure that he or she returns to purchase your products? If the customer is looking for a way to increase their reach through social media, you should see how the customer can achieve this.

How to Use the Lean Analytics Canvas

It is a good idea to use a lean analytics canvas when you want to understand or view some data. Look at how you can use this canvas:

- Choose the required result or the outcome for the project

- Choose the dates against which you can meet a goal or an objective, and set the goal against that date

- Make the list of functionalities and actions that you will need to complete to achieve this goal

- Test the functionality one at a time and implement that functionality

- Measure the impact of the functionality on the process

- Proceed to the next box once you meet the desired goal

PART THREE: Agile Project Management

Chapter Eleven: What Is The Agile Framework?

The Agile framework, unlike other project management methodologies or frameworks, focuses on iterating quickly and satisfying the demands of customers. The Agile framework is based on the Agile Manifesto, which will be looked at later in this chapter. The agile framework can either be called a methodology or a process. One of the values in the Agile manifesto states that the philosophy prioritizes interactions and individuals over tools and processes. Most agile teams use these frameworks as the starting point, and they customize the elements in the methodology to meet their needs.

Organizations use different agile frameworks, and most of them modify some parts of the framework as per their needs and build their agile processes. Now look at some of the common agile frameworks in the industry:

- Scrum

- Adaptive Software Development (ASD)

- Extreme Programming (XP)

- Scaled Agile Framework (SAFe)

- The Crystal Method

- Rapid Application Development (RAD)

- Dynamic Systems Development Method (DDSM)

- Disciplined Agile (DA)

- Lean Software Development (LSD)

- Feature Driven Development (FDD)

Which Framework is Best?

Since there are numerous agile processes in the market, it is hard to choose the right one. That being said, there is no CORRECT agile process that you can choose from. You must consider different factors about your business before you choose the framework that will work best for you. Some of these factors include:

- Structure/size of your product portfolio

- Company size

- Needs of stakeholders

- Available resources

- Team structure

Every framework has its pros and cons, and a framework that works for one business does not necessarily work for your business. You should experiment and try the frameworks before you identify the framework that works best for you.

Scrum, Extreme Programming and Kanban

Now look at some of the common agile frameworks in the industry: Scrum and Kanban. Scrum is an agile framework that gives an organization the capability to manage and control incremental and iterative projects of different types. Two other agile frameworks that are accepted by the industry are Kanban and Extreme Programming.

The core principles of extreme programming are engineering principles. Organizations following this approach focus primarily on the delivery of high-quality software. A team using this framework or methodology collaborates and works in short, flexible development cycles, and they are eager to adapt and change the methods used in the development of the output. In Extreme programming, teams work on small user stories and plan small releases or prototypes of the output to obtain feedback from the clients or stakeholders.

Kanban is another agile methodology that focuses on the visualized workflow. In this method, the work is broken down into smaller sections or pieces that can easily be managed by the team. The Kanban methodology helps organizations and project managers identify the waste or bottlenecks in their processes. It also helps teams reduce the wait time between the work done and the delivery of the product. Kanban can do this since the methodology adheres to strict process policies and helps the teams manage and measure the flow of work.

The Agile Manifesto

Scrum is a framework and not a mathematical process or methodology. You still need to think and make choices. One of the biggest advantages of the Scrum framework is that you can make discretionary decisions that are best for you, based on the feedback you receive from your customers.

In 2001, seventeen software and project experts, who were successful in their divergent processes, agreed upon the following values that best suited their programming methodologies.

They believed that they uncovered better ways of developing software and wanted to help people all over the world do it better, too. These values formed the Agile Manifesto, and any project management tool that uses agile must adhere to these values:

- Interactions and individuals take precedence over tools and processes

- Working software is more important than comprehensive documentation

- Collaborate with customers before you begin any negotiations

- Always respond to change and tweak your plans whenever necessary

This means that the team should be concerned more about the items on the left in the points mentioned above.

Even though the Agile Manifesto and principles were written by and for software experts, the values remain valid for any Scrum project you embark upon. Just like GPS was designed by and for the military, it does not mean that people cannot benefit from it when they sit in their car and head towards a new part of town. For more information on the history of the agile manifesto and its founders, visit http://agilemanifesto.org.

Agile Principles

The founders of Agile did not stop only at the values. They also defined twelve principles to expand on those values. You can use these values in your Scrum project to check if your framework adheres to the goals of agile:

- It is important to satisfy the customer by delivering the product or software early. The alternative approach is to deliver smaller sections of the product to the customers.

- Ask your customers for feedback and make changes to the product, even if it is late in the development stage. Every agile process allows you to make some changes to the product to improve customer satisfaction.

- Deliver working prototypes of the software at regular intervals to the customer.

- Developers and business people should work together throughout the project.

- You should give the developing team all the support it needs to ensure that the job is done.

- Have regular meetings where you can convey information to the team effectively and efficiently.

- You can measure your progress by preparing working prototypes of the software.

- Developers, sponsors, and users should maintain a constant pace throughout the project.

- Good design and attention to technical detail enhance agility.

- It is essential to keep the process simple. This means that you should maximize the amount of work your team does not have to do.

- Self-organizing teams provide the best designs and architecture.

- The team should reflect on what it should do better to become more effective and then adjust its behavior accordingly.

The principles do not change, but the tools and techniques to achieve them can.

Some of the principles will be easier to implement than others. Consider, for example, principle two. Maybe your company (or group or family) is open to change and new ideas. To them, Scrum is natural, and they are ready to get started. But on the other hand, some may be more resistant to change.

How about principle six? Is working face to face possible in your project? With the Internet and globalization of workforces, you may have team members from India to Russia to the United States of America. Instead of worrying about how this principle cannot apply to your team, you should identify a solution. Can you use Hangouts or Skype to stay in touch with your team? Do you prefer a teleconference? This is not the intention of the sixth principle, but if you are to improve tomorrow, you should focus on how to deal with today. This means that you should learn to adapt to change.

You are bound to have unique challenges. Do not let a hiccup or less-than-perfect scenario stop your team from working on the project. Part of the fun in using Scrum is to work through issues and get to the results. The same goes for the twelve principles. If you adhere to the principles listed above, you can improve your team's efficiency and quality.

Platinum Principles

Experts will suggest that you use these principles when you work on a project since they improve efficiency and assist in the implementation of the process.

Visualize Instead of Writing

Overall, people are visual. They think pictorially and remember pictorially. For those old enough to remember encyclopedias, which part did you like best? Most kids like pictures and adults are no different. They are still more likely to read a magazine flipping first through images, and then sometimes going back for articles that piqued their interest (if at all).

Pictures, diagrams, and graphs instantly relay information. However, if you write out a report, people will stop reading if there are no diagrams to support the claims made in the report.

Twitter was interested in studying the effectiveness of tweets with photos versus those that were text only. It conducted a study using SHIFT Media Manager and came up with some interesting results. Users engaged five times more frequently when tweets included photos as opposed to text-only tweets. The rate of retweets and replies with photos doubled. However, the cost per engagement of photo tweets was half that of text-only tweets.

When possible, encourage your team to present information visually, even if that means sketching a diagram on a whiteboard. If anybody does not understand it, they can ask, and changes can be made immediately. Also, with technology today, you can make simple graphs, charts, and models easily.

Think and Act as a Team

The heart of Scrum is working as a team; however, the team environment can, at first, be unsettling because, in the US corporate culture, the opposite is encouraged—an individual competes with his or her peers. "How well can I succeed in this environment so that I stand out and get the next promotion?"

In Scrum, the project survives or dies at the team level. By leveraging the individual's talent to that of a team, you take the road from average to hyper-productive. According to Aristotle, "The whole is greater than the sum of its parts."

How do you create this team culture? The Scrum framework itself emphasizes the team. Physical space, common goals, and collective ownership all scream team. Then add the following to your Scrum frame:

- Eliminate work titles. No one "owns" areas of development. Skills and contribution establish status.

- Pair team members to enhance cross-functionality and front-load quality assurance, then switch the pairings often.

- Always report with team metrics, not an individual or pairing metrics.

Avoiding Formality

Have you ever seen a knockout PowerPoint presentation and wondered how much time someone spent putting it together?

You should never think about doing this for a Scrum project since you will be wasting too much time. Instead, you can scribble it on a flip chart in 1/1000th of the time and stick it up on a wall where people will look at it and then get back to creating value. If it requires discussion, walk over to the concerned parties and ask them now or whenever the need arises. Focus your valuable time and effort on the product instead of prepared presentations.

Atos Origin produced independent research showing that the average corporate employee spends close to 40 percent of his or her working day on internal emails that do not add any value to the business. This means that the real workweek does not start until Wednesday.

Pageantry is too often mistaken for professionalism and progress. In Scrum projects, you are encouraged to communicate immediately, directly, and informally whenever you have a question. You also save time since you work closely with the other members of the team. You should identify the simplest way you can get what you need with the goal of delivering the highest-quality product in mind.

Before long, your projects will evolve a Scrum culture. As people become educated on the process and see the improved results, their buy-in for barely enough will increase accordingly. So, bear through any initial push back with education, patience, and consistency.

Chapter Twelve: Start An Agile Project

Most traditional project management methodologies use a linear process. Here, you will understand the project, plan the tasks, decide on a strategy, and build the right solution to cater to the problem before you move it to production. You will then fix any problems in that solution using data from regular assessments. This is also known as the waterfall approach since it includes cascading steps. Traditional project management methods do have issues with the timeline and budget since conventional development sequences do not allow for any changes in the requirement. These sequences do not allow the teams to make any changes to the demands of the clients or budget.

Understanding Agile Project Management

In the previous chapter, you learned more about what the agile framework is. You know that the agile framework does not focus on fixed sequences, but works in cycles that facilitate continuous collaboration, improvement, and innovation. The customer or client is always involved in the project. Now look at what agile project management is all about.

Understand the Problem

It is important to understand what your customers or clients need. To do this, you should ask them about the different problems that they are facing, and what the problem statement is. As an agile project manager, it is important to focus on the end-user. You may need to conduct thorough research of the market or conduct interviews with the customers to answer these questions. In simple words, you should focus on trying to understand how to measure your success.

Assemble the Right Team

Now that you have understood the problem, you should focus on trying to set up a team that has the necessary experience and skills to solve the problem. This may mean that you need to bring in people from different departments or hire external consultants. In some situations, you will also need to develop the skills of the existing team members.

Brainstorm

The team can now work on developing ideas on how to solve the problem. At this point, every idea should be looked at. Ensure that you encourage every member of the team to innovate.

Build an Initial Prototype

Once you have identified the potential solution, you should work on developing a prototype. This will not take too much time, and you should remember that agile is a project management methodology based on creativity and flexibility. Once you build the prototype, let the customers look at the prototype so they can give you their opinion or feedback. If the feedback is negative, inform the team that they will need to work on a different model which aligns to the customers' requirement. This is one of the best approaches since you learn the issues with the design very early, which gives you sufficient time to make the required changes. You will not risk working for days or months on a prototype, and realize at the end of the project that this is not a good fit for your client.

Decide the Boundaries

Based on the feedback you receive from the clients, you should decide the scope of the project. You can add or remove features based on the feedback you receive for the prototype. You can maintain a document that outlines the scope of the project, and update that document whenever necessary. You should also update that document as the project progresses.

Plan the Major Milestones Using A Roadmap

The next step is to set the milestones that you and your team must meet when developing the product. You do not have to add too many details to this roadmap. In fact, it is important that the milestones you set are easy to meet and flexible. You can, however, decide the different components that you will use to make that product and ensure that you meet the deadlines.

It is not only the features of the product that you should look at, but you should also look at the goals. For example, if your client wants to build on their customer base, you should understand how they plan to achieve that goal. Ensure that you know exactly what needs to be done to meet your customers' demands.

Plan Sprints

Sprints are short development cycles that last anywhere between one and four weeks. If you want to ensure that the development rate is stable, you should maintain the same length for every sprint. When you plan a sprint, you should devise the list of tasks that the team should complete and choose realistic targets.

The goal is to assemble a product that is functional in the shortest time frame. Your team can then work on improving that product in the subsequent sprints. You can ensure that the product is developed successfully only when every member of the team collaborates and cooperates. People in the team should always be given a chance to express their concerns or opinions.

Check In Every Day

A daily stand-up or meeting will make it easier for you to identify any issues early in the project. A stand-up is a small meeting, around fifteen minutes long, where all the members of the team will talk about their progress. Every team member should tell you what he or she worked on for the previous day, what they want to achieve today, and whether there are any concerns or problems that they have identified. It is your responsibility as the project manager to keep the team on track and ensure that you work with them to resolve any issues.

Review the Sprint

You should sit down with your team when the sprint ends and evaluate the progress of the team. Ask them what they thought they did well, the lessons they learned, the parts of the task that they can improve, etc. It is important to review and have meetings with your team daily to monitor the health of the project.

You should also ask your team to give you real-time project updates and ensure that you always keep the clients in the loop. Your clients should also be given the freedom to check up on the project at any time and give you feedback. The client should be made aware that the team is working on the project. When you interact with your clients regularly, you can iron out any bugs.

Plan the Next Sprint

You should always use the sprint system until you complete your project. Ensure that you are open to change and always make changes to the process based on the end-user or client feedback. You must always strive for excellence and never forget about the design. If you follow the agile project management methodology, you need to rely on both technical expertise and strong interpersonal connections. You must choose an effective communication channel and ensure that the entire team uses it.

Completion and Release

Once you develop the product, and you receive positive feedback from the end-users and clients, you can manufacture that product and release it into the market. The agile project approach, however, does not end here. You will need to make some adjustments to the product if there are any defects or bugs in it.

You will learn to communicate better with every project. It may be hard to cope with agile project management, at first, but you will soon begin to wonder why you ever used traditional project management methods.

Chapter Thirteen: Agile Versus Scrum Versus Kanban

Over the years, project management has significantly evolved, and numerous project management tools help to make those changes to project management. If there is someone following a project management trend, they will know that technology is only a small part of that discussion. Agile, Kanban and Scrum, and other project management methodologies dominate the discussion. In this chapter, you will understand these terms and look at the difference between these methods.

Differences Between Agile, Kanban and Scrum

Agile is a project management methodology used to break the tasks in a complex project into smaller chunks of work that can be managed easily. Agile project management was earlier used in software development projects to improve the speed at which the project is completed, but it is now being used in different industries. Agile is a set of principles that were defined in the Agile manifesto, and these have been defined in the previous chapters. Kanban and Scrum are two methodologies that follow Agile principles. In simple words, if you want to implement an agile framework in your business, you can use Scrum or Kanban to do this. This definition was proposed by Nicholas Carrier, Associate Partner at London-based Prophet.

Kanban and Scrum are agile project management methodologies that have some differences. Carrier noted that these methods also have some similarities. Each of these methods uses a board that shows the project manager the status of the tasks where people will move between the three categories:

    1. Tasks that have not been started

    2. Tasks that are in progress

    3. Tasks that are complete

Differences Between Scrum and Kanban

Both Scrum and Kanban share some similar traits, and people make the incorrect assumption that both Scrum and Kanban are two sides of the same coin. This is far from the truth since these agile methodologies are very different. The Scrum methodology will break the development cycle time into work periods with time limits called sprints. These sprints last for a maximum of two weeks. Jessica D'Amato, who is a project manager at Dragon Army, said that a project manager could plan the initiatives that should be completed by the team within a two-week sprint. They should also hold meetings to understand the progress and how the task is moving along. A project manager also uses this meeting to demo any new releases to the client before they launch it. There are three prescribed roles in Scrum:

Product Owner

The product owner focuses on the initial planning of the tasks. He or she also works on prioritizing the task and communicating with every member of the team.

Scrum Master

The Scrum Master is responsible for assessing the status of every process during a two-week sprint.

Team Members

A team member is an individual who will carry the tasks out in a two-week sprint.

Scrum follows a pre-defined structure or framework while Kanban does not. Kanban is a methodology based only on the list of items or tasks that are in the backlog. There is no set time within which a task on the Kanban board needs to be completed, but every task on the board is given a priority. This board has different columns, which make it easier for the project manager to know the status of any task in the project. They will know which task is currently being worked on and which task is completed.

Joe Garner, who is the project manager in a computer design firm, mentioned that Kanban focuses on improving every aspect of the process. Kanban is a method that can be used to manage the creation of any product with the vision of delivering the output. This project management methodology can enhance the processes being followed in a company by improving them without making a change to the entire system.

Agile Pros and Cons

Garner talks about how agile methodologies like Kanban and Scrum can provide an iterative and incremental approach to complete any project when compared to traditional project management methods where a linear approach is followed. He mentioned that Agile focuses on the different business requirements and also creates the product, which will need to be delivered to the customer. This product is released in small units. The Agile framework focuses on accountability, transparency, and strong teamwork. This will ensure that the product aligns with the goals of the company and the client.

As Garner mentioned, agile management gives the teams the flexibility to improve their processes continuously. Having said that, this could lead to some delay in the final delivery date. This problem often arises in a digital transformation team since the executives believe the agile methodology, but they do not have the necessary resources to work iteratively on a process and improve it. When this happens, a team may be under too much stress to complete work on time. They will then switch to the old methods, which will result in the delivery of a product with low quality.

Scrum Pros and Cons

The Delivery Manager at Oak Brook, Brijmohan Bhavsar, talks about how Scrum provides high visibility and transparency of the projects. It also allows the team to accommodate changes to the tasks or processes with ease. In addition to this, Garner also mentioned that Scrum helps to define the roles of every team member. It also enables better collaboration, which will ensure that a project is completed faster. Carrier mentioned that they use Scrum in strategy projects as well since this enables them to communicate with business stakeholders from different departments like technology, marketing, and operations frequently. It allows the teams to collaborate and make the right decisions to obtain the required outcome. Bhavsar also mentioned that breaking a complex task into smaller chunks of work could lead to creating a poorly defined task that could affect the scope of the project.

Kanban Pros and Cons

Kanban is a model used to present any change using additional improvements. This methodology gives a visual of what the team is currently doing. The Kanban board plays an important role in displaying the tasks and workflow. Bhavsar believes that the board also assists with optimizing the tasks that each team must perform. Carrier, however, mentioned that the Kanban methodology could lead to poor productivity since it lacks a structured framework. Kanban does not focus on a cross-functional team since it does not use sprints. He believes that a sprint will help to assign the time a team can take to perform the task. A sprint will drive the team to work towards delivering the process at an increased speed. This is important for digital transformation.

Which One Should You Choose?

You should look at the business requirements before you make the decision between which agile methodology to implement or whether you should implement the agile framework in your business. Bhavsar states that it is best to make this decision based on whether you want the project to be completed faster or if you want to improve the process. He mentioned that it is a good idea to use Scrum if you only want to work on producing work faster. If you want to improve your processes, you should use Kanban. If you want to use a linear workflow, you should implement a waterfall model.

Chapter Fourteen: Step-By-Step Scrum

Scrum is a project management tool that allows teams to use their experiences to make decisions about the project. This tool is a great way for teams to organize their projects, regardless of their size. Through Scrum, a team can identify if the process or approach it plans to use to meet the objective or vision will generate the intended results.

If there is a task that you need to complete, you can use Scrum to provide some structure to your approach. This will help to increase efficiency and improve results. Scrum breaks the tasks down into manageable pieces and prioritizes those tasks depending on what the customer demands. This will help you identify the tasks that you need to complete today, tomorrow, and the next day. You can also see how well you are progressing and where you should adjust to counter any inefficiencies in your approach. This will help you improve your speed and efficiency.

Arts used the concept of empirical exposure modeling since the beginning of time. For example, a sculptor will use a chisel on the material he or she is using to make a sculpture and will check the results before he makes any adaptations. We now use this concept to work on developing software. In this method, the developer does not use a simulation to understand how the project will progress. Instead, he observes actual results and learns from those experiences to develop the project.

Basics of Scrum

A scrum is a tool that helps you devise a process that you can use to develop or manage a task. It provides a framework that will allow you to define the roles of every team member and the work that each member must complete. You can use this framework to prioritize your work and become more efficient in completing the assigned task. Frameworks are not as prescriptive as methodologies since they allow you to add structures, tools, and processes that will complement the primary task. Through this approach, you can observe the process and include other processes that work well with the basic processes wherever necessary to enhance the process. You can complete a task in a few hours or weeks using this framework.

Using Scrum, you can improve the performance of your team at a nascent stage since this project management methodology uses a gradual and repetitive method. When you use the Scrum project management methodology, you should let your team members choose the tools that they can use to improve their performance or conclude the process. There will be no hierarchy in the management, which will reduce the number of progress reports and redundant meetings. If the objective is to get the job done, you should use Scrum since it will help you improve your process.

Scrum is a term used in rugby. The teams form a Scrum or huddle where the forwards from one team interlock their arms and push against the forwards of the opposing team who also have their arms interlocked. The referee then throws the ball into the middle of this huddle. The players work as a team to move the ball down the field. Scrum is like rugby in the sense that it relies on people from different domains with different responsibilities to work towards a common goal.

People believe that only processes in IT, software development, or other technical processes can use Scrum. What they are unaware of is that you can use Scrum to improve any process regardless of whether it is small, large, personal, social, or artistic.

The Roadmap to Value

There are some techniques that you can apply as an extension to Scrum. It is important to remember that these extensions cannot replace Scrum but only improve its function. This book only recommends the practices that experts across the globe have tried and tested. You can decide if you want to use these extensions to improve Scrum, depending on your project.

The aggregation of the common practices of Scrum is known as the roadmap to value. This map consists of seven stages that help you define the vision of your project all the way through the process. In other words, these stages will help you understand what you want to achieve at every stage of the cycle.

Once you define the vision, you can break it into smaller segments to see how you can achieve it. You can then develop a cycle that is efficient and provides results every day, week, and month. The stages are as follows:

- Vision

- Product Roadmap

- Release Planning

- Sprint Planning

- Daily Scrum

- Sprint Review

- Sprint Retrospective

If there is an idea lurking at the back of your mind for years, you can finally put it into action using these seven stages. You will learn more about these stages throughout the book. These stages give you information about the feasibility of your project. You will also learn which parts of the process need improvement.

A Simple Overview

Scrum is a circular and simple process that allows you to inspect and adapt to changes in the process constantly. This section provides an overview of the Scrum process.

Product Backlog

Scrum breaks the process into smaller pieces of work and creates a to-do list for the team. This list is known as the product backlog, and you should follow this list to ensure that your process is efficient and flawless.

Prioritize

Scrum then labels the tasks on the list based on priority. The team should ensure that it completes tasks with a high-priority within a specific period. This period is known as a sprint, and it defines the amount of time a member should spend on the task.

Scrum allows you to quickly adapt to technology, constraints, new innovations, regulations, and market forces. The objective is to work on tasks with high priority and complete them within the prescribed time. All high-priority items go through the following steps:

- Defining the requirement

- Designing

- Developing the requirement

- Testing the process

- Integrating feedback into the process

- Documenting the process

- Confirmation of approval from the stakeholders

Every task on the list will go through the steps mentioned above, regardless of how big or small the task is.

When Scrum decides that a process or task is shippable, you can release it to the customers and take their feedback. You do not have to worry about whether the end-users will like the product or not. Instead, you can work on high-priority tasks and show the stakeholders some tangible results. You can also ask them for feedback and use that feedback to improve the process or generate a new task. During the Scrum meetings, the development team and product owner will add these tasks to the product backlog. You can then prioritize these tasks against the existing ones in the product backlog.

As a process developer, you know that effectiveness is more important when compared to efficiency. You will learn to be efficient if you have an effective team or process. As a team, you should focus on high-priority tasks first. Management consultant, educator, and author, Peter F. Drucker rightly said, "There is nothing as useless as doing efficiently that which should not be done at all."

When the team completes a task, it can share the results with the stakeholders. There are times when the stakeholders will give you some feedback about the process. You should learn to incorporate that feedback into the process and work on the requirement again. Ultimately, you will develop a process that is effective. You will also learn to deliver the results faster and with better quality.

Teams

Regardless of the scope of your project, your team will have similar characteristics. The size of the team will vary, but the roles of the members of the team will remain the same. This book will cover the roles of each member in detail.

The heart of every team is the development team, and these members work together to develop the product. The development team works closely with the Scrum master and the product owner. The roles of both the Scrum master and product owner will be detailed, but in simple words, their roles align the development process to the needs of the stakeholders. They also help to eliminate any distractions for the development team, which allows the team to focus only on their job—developing! The Scrum team is accountable for every task in the cycle. As a team, the members should identify the most effective way to achieve their objectives regardless of the environment they are in.

Stakeholders are not a part of the Scrum team, but it is important to include them since their feedback will impact your project. These stakeholders can be internal or external and include customers, the marketing team, investors, and the legal team.

Governance

The product owner, development team, and scrum master are the three independent units of a scrum team. These units should work together to improve the efficiency of the processes and the team.

Product Owner

The scrum cycle defines the different requirements that the team has to meet or develop. It does not look at the amount of work that the team needs to complete at the end of every sprint. For instance, if the team is working towards developing software, the objective that the team may have set is to develop a prototype at the end of the sprint where the user can easily be moved from the landing page onto the main website. The product owner will need to speak to the stakeholders to understand what the requirements are and prioritize those requirements. They will then need to pass this information on to the development team.

Development Team

The development team will always look at the features that the product manager has listed after having a discussion with the stakeholder. The team can then decide the tasks that they can address or complete in a sprint.

Scrum Master

The scrum master can choose the different processes the team must follow to develop the final product.

The developers of Scrum did not simply assign these roles, but they used their project management experience to establish these roles and define them. The developers are aware of different types of teams and know what works best for the team. It is important to ensure that Scrum teams only have full-time resources. When you share resources between processes, they will tire out and will stop working to their full potential. Have you ever heard of a football team having part-time players? If there is such a team, it will not be a successful one.

Scrum Framework

As mentioned earlier, Scrum is not just a methodology but also a framework. This framework allows every member of the team to understand their responsibilities. It also gives the team a chance to inspect the different elements in the cycle and include some new elements to the process if required. They can also include new processes to the product backlog based on feedback. This framework also provides some processes and tools that teams can use to meet their timelines.

Scrum follows a 3-3-5 framework.

- Three artifacts

- Three roles

- Five events

These elements fit into the cycle. Since the scrum framework follows an iterative and incremental cycle, you can learn new ways to improve that process using the Scrum cycle. The scrum framework is quite straightforward.

Artifacts

- Sprint Backlog
- Product Backlog
- Product Increment

Roles

- Development Team
- Product Owner
- Scrum Master

Events

- Daily Scrum
- Sprint
- Sprint Planning
- Sprint Review
- Sprint Retrospective

The artifacts in the above list are related to the work or tasks that the teams should finish. These artifacts also include the requirements or the demands of the end-users or customers, which can be completed in one scrum cycle. It is important for the team to review the artifacts during the scrum cycle and ensure that the team is working towards completing or meeting the objective.

Every artifact, event, and role have an important purpose in the scrum. Use the roadmap mentioned above to complete any task or project. It is important to remember that Scrum is merely a framework that helps you see what you are doing and where you are in the process. You can choose from the different techniques or tools that you think are necessary to complete any project. Scrum does not define the processes that a team should follow to meet the objective.

One can understand scrum easily, but it is slightly difficult to implement. How to set up a scrum team will be detailed later in this chapter.

Feedback

Scrum, unlike other project management methodologies, allows the team to obtain feedback from the end-users or stakeholders whenever the team completes a task in the backlog. The team will, therefore, learn more about the processes and see which ones are working for them and which need improvement. The teams can also identify the areas where they should enhance efficiency.

The feedback loop works in the following manner:

- The feedback is shared daily between the members of the teams since they work on different tasks in the sprint

- The product owner and the development share feedback every day

- The feedback from the product owner when the end-user accepts or rejects any requirement

- When the sprint is complete, the team will receive some feedback from the business

- The marketplace will provide some feedback to the development team when the product is released into the market

A team can learn more about the processes if it follows the scrum methodology when compared to the traditional project management methods. The latter focuses only on the development of the artifact, while the former only focuses on continuous development. Since you receive regular feedback from the stakeholders about the requirement, you will know what changes to make to the process at an early stage to ensure that you release the final product into the market. At the end of the project, the team does not have to worry about how the customers or end-users perceive their products because you have been communicating and receiving feedback from them all along the way.

Steps to Follow

Here are some steps that you can use to implement Scrum at your workplace.

Define the Scrum Team

A scrum team usually has a minimum of five members and a maximum of eight. These members have a combination of capabilities and can include people from different departments. The members of the team will work together. This team is responsible for delivering the product requirements at the end of every sprint.

Define the Sprint Length

The sprint is a time-box, which can last anywhere between seven and 30 days, and the length of the sprint will remain the same across the entire project. The team will have a planning meeting before every sprint. It is during this meeting that the work for the sprint is planned, and the team will commit to completing this work by the end of the sprint. At the end of the sprint, a meeting is held to review the work completed during the sprint. It is during this meeting that the improvements are reviewed, and the next sprint is planned. If you do not know how long the sprint should be, you can start with a minimum of two weeks.

Appoint the Scrum Master

The Scrum master is the most important person of the scrum team. A scrum master should ensure that the group is effective and progresses well. In the event that there is an issue in the project, the scrum master will work on resolving that issue for the team. The Scrum master is the project manager for the scrum team, except that he or she cannot decide what task the team can work on. This person should avoid micro-managing the team. The scrum master will help the team plan the workflow for the project.

Appoint the Product Owner

A product owner is a person who will be in charge of the team, and he or she must ensure that the team always produces some value to the business, stakeholder, client, or the end-user. The product owner will write the requirements of the clients in the form of a story and will prioritize that story and add it to the product backlog.

Create the Initial Backlog

The product backlog contains a wish list of the requirements or user stories, and it is expected that you complete these tasks in the project. An important story or requirement should always be present at the top of the backlog list. The list of tasks in the backlog should always be ranked based on their importance. A backlog will contain two types of items:

Epics

A high-level story or requirement has a very rough sketch. There are very few details about the story.

Stories

A story is a more detailed requirement that includes the different tasks that should be completed. You can break an epic into numerous stories. You can also break a story down into smaller tasks that a team can work on. Every story can be a different type, including defect/bug, development, chore, etc. Any member of the team can write a new story and add that to the product backlog at any time.

When you go further into the backlog and look at the items at the bottom of the list, these requirements will be epics because they do not have too much detail. An epic or story will rise in priority, depending on the details it contains. These details make it easier for the teams to work on the tasks. A product owner can re-prioritize the backlog as they see fit and at any time during the project.

Plan the Start of The First Sprint

Based on how the team prioritizes the tasks in the backlog, the members can pick the items on that list. The usual process followed is to choose the task with the highest priority. The team should then brainstorm and decide on how much of the task in the backlog they can complete before the next sprint. When the team agrees on the decision made, the sprint will start. The team should now work on developing the stories.

Close the Current Sprint and Start the Next One

When the team finishes a sprint, it is important to ensure that the team completes all the work planned for the sprint. If this is not the case, the team should decide if the work that is left is moved to the next sprint or whether it should move to the backlog.

The team will now review and discuss what processes worked well during the sprint and also identify the processes that should be improved in the next sprint. The team should then discuss and plan the next sprint. This process is repeated until the final sprint is completed. There is no limit on the number of sprints that a team can have in a project, except if there is a strict deadline or if the entire backlog is complete. If these criteria are not met, the sprint will continue indefinitely.

Chapter Fifteen: Create A Kanban Project

As mentioned earlier, businesses use the Kanban methodology to manage processes or workflows while the tasks are in progress. To do this, businesses use Kanban boards that are present in the management system. A Kanban board can be used to visualize the workflow and understand the stage of every task in the workflow. The objective of this system is to identify the potential bottlenecks or impediments in the tasks and identify a solution or fix the issue.

A Brief History

Kanban was developed by the Japanese Industrial Engineer and Businessman for Toyota automotive, Taiichi Ohno, in the 1940s. He created the Kanban system to help the various teams involved in the manufacture of a car to plan their work. It also helped these teams identify any potential issues and develop a solution for those issues. The Kanban software aimed to manage the workflow at every stage of the process.

Taiichi Ohno developed the Kanban methodology because the business was not functioning well during the 1940s when compared to the competitors in America. Kanban helped Toyota control the systems involved in the production of their automobiles, which improved productivity. The Kanban method also helped the business reduce the amount of raw material that went to waste.

A Kanban system controls the value chain. It starts with the supplier of raw material and ends at the customer. This makes it easier for teams to avoid any disruptions in the supply chain functioning, and also reduces the excess of any inventory at different steps of the process. The Kanban method requires constant monitoring, and the project manager must identify any bottlenecks in the system. Since most companies want to achieve high output with low delivery times, they should choose the Kanban project management methodology since that helps to improve efficiency.

What is the Kanban Method?

As mentioned earlier, the Kanban method was developed by Taiichi Ohno to improve the manufacturing process in Toyota. David Anderson used this methodology in the IT industry for software development in 2004. He then went on to define the method using concepts like queuing theory, pull systems, and flow using the work by Eli Goldratt, Peter Drucker, Edward Demmings, Taiichi Ohno, and others.

Kanban Change Management Principles

The Kanban methodology uses a list of principles and practices that help to manage or improve the workflow. This method is evolutionary and non-disruptive and helps to improve any process being performed in the organization or any project. You, as a project manager, can implement the principles and practices of Kanban to improve every step in the business process. You can improve the workflow by reducing the time spent on completing the task, increasing customer satisfaction, and maximizing every team member's capabilities. Now, look at some of the foundational principles and practices of the Kanban methodology.

Foundational Principles

Always start with what you are doing now

When you implement the Kanban methodology, you do not have to make any changes to the processes immediately. You should, instead, focus on the current workflow and look for the sections where you can make some improvements wherever needed. You must also ensure that your team members accept the changes that you make to the process.

Pursue Evolutionary and Incremental Change

The Kanban methodology allows every member of the project team to make some changes to the process. Since they work together with the project manager to make these changes, there will be very little resistance from the team and the organization.

Respect Current Roles, Responsibilities, and Designations

Unlike other project management methodologies, the Kanban methodology does not require the organization to change the structure. The business does not have to make any changes to existing functions and roles that are not performing well. The only processes that must be changed, if necessary, are the tasks or processes within the project.

Encourage Acts of Leadership

The Kanban methodology encourages continuous improvement to processes at every level in the organization. People at any level are allowed to lead the change or identify some improvements to the processes. They are always encouraged to identify ways to improve the processes followed to meet the objectives of any project.

Core Practices

Visualize the Workflow

When you begin to implement the Kanban methodology, you should visualize the workflow of the tasks. You should visualize the steps in the workflow on the Kanban board, so you know the steps that must be followed to complete the task. The Kanban board can either be simple or complex based on the different types of items or tasks in the workflow that the team should deliver.

Once the team and the project manager visualize the process, it is important to visualize the current work that is being done by the team. The team can do this by using different cards or colors to identify the different classes of work. You should also include different colors and columns to define the tasks or processes, which the team works on. It is a good idea to use a Kanban board, which can help to redesign the process.

Limit or Reduce Work-in-Progress

It is important to limit work-in-progress to implement the Kanban system. This will encourage your team to complete work that is ongoing before taking up any new project. It means that the team can only take up a new process when work that is in progress is marked complete. This increases the team's capacity to bring in more work.

It may not be easy to identify the limit of work-in-progress. You may begin with no WIP limits, and author Donald G. Reinertsen had suggested that a team should start with no WIP limits and then observe how the team performs when you begin to use Kanban. When you have sufficient data, you can define your WIP limits at every stage of the workflow. Most teams start with a WIP limit that is either 1 or 1.5 times the number of people in the team working on a specific task.

It is beneficial to the team to limit WIP and put WIP limits since the team members will finish their tasks before they move on to new tasks. This also communicates to the stakeholders and customers that there is a limited capacity of work that can be performed by the team. Therefore, careful planning must be involved when any new task or request is made to the team.

Manage Flow

Once you implement the first two activities on this list, you should work on managing and improving the workflow. The Kanban methodology will help the project manager handle the workflow by splitting the workflow into stages and also understand the status of every task. The WIP limit will be set based on the tasks in the workflow. Every member of the team will work on completing the tasks within the WIP limits set since they know that the work will be piled up if there is any task that is held up. This will affect the speed at which the work can be delivered.

The Kanban methodology will help the project manager and the team analyze the system, and also adjust some tasks in the workflow. It will also allow the team to understand how they can spend time efficiently to complete a task. It is important to look at the intermediate stages to observe the tasks, identify bottlenecks, and resolve those bottlenecks in the workflow. Teams should always analyze the time that a task stays in the intermediate stage and identify some ways to reduce the time spent on those stages. This is an important step that one should consider if the team wants to reduce the time spent on the project.

Once the workflow improves, the team will learn to deliver any project smoothly. When the team is more predictable, they will become more comfortable to realize the commitments that are made to the customers and also complete the work within the set timelines. Teams should always forecast completion times, which will help them improve the functionality of the team.

Make Policies Explicit

The team should work on defining the policies explicitly. They should also find a way to visualize those policies so they can explain the work that is being done in the process. A project manager should always create the policy and define the amount of work the team should complete. The policies can always be written at a broad level. These policies can also include some checklists, which must be ticked off at the end of every task to ensure that the team is meeting all the required criteria. These lists should be made for every task that the team performs to ensure that the team maintains the workflow. For instance, the project manager should include the definition of when the task is complete or understand whether a task is pushed or pulled in the policies.

Feedback Loops

It is important to include feedback loops in any system, and the Kanban methodology requires every member of the team to implement a constructive feedback loop. The team, as a whole, should look at the different stages in the process on the board. They can then use the information on the board to generate metrics and reports that can be used to improve the process.

Most teams do not understand the mantra "Fail fast and fail often," and this may not work for most teams. Having said that, this method would make it easier for the project manager and the team to identify any issues at the start so that the process has fewer errors. It is important for every team to have a feedback loop in the process.

Improve and Evolve Collaboratively and Experimentally

The Kanban methodology is an improvement process that allows teams and people to make changes to the processes before they make a huge improvement. This will make it easier for teams to implement the changes quickly. The Kanban methodology allows members to use statistical methods where the members can build a hypothesis and test that hypothesis to understand that outcome.

Teams need to evaluate a process and then improve the process wherever necessary. It is important to observe these changed processes and assess the impact of that change on the process using the Kanban board. These results will help the project manager and teams evaluate if the change made to the process is improving the process or not, and decide to keep the change or remove it.

The Kanban methodology helps a team collect all the information required to assess the performance of the process and also every member of the team. This data makes it easier for the project manager to generate the necessary metrics to evaluate the performance and tweak any other processes if needed.

Implementing Kanban

When it comes to implementing Kanban, you must ensure that you have enough patience since the implementation of Kanban is a continuous process. That said, the results will never disappoint you. Most project managers use the steps mentioned below to implement the Kanban methodology in their company or project.

Step One: Visualization Of Workflow

This is the most important step that you will need to perform when you implement the Kanban methodology. Remember that the Kanban method is primarily based on visualization. Therefore, you will need to create the Kanban task board first when you choose to implement this methodology. You can either use a physical or digital board since there is no difference between the two. The principle of the board is the same. This board should represent the stages or the status of every task like "to do", "in progress", or "done". The tasks should then be placed under these statuses depending on the stages.

Step Two: Limit the Amount of WIP

As mentioned earlier, one of the main principles of Kanban is to limit the number of tasks in the work-in-progress stage. You must limit the number of tasks that you work on when you implement Kanban, which will make it easier for you to spend your time efficiently. Some believe that it is a good idea to handle multiple processes or tasks at once, but this is not the case in Kanban. You can only use Kanban if you limit or reduce the number of units or tasks at the work-in-progress stage.

Step Three: Switch to Explicit Policies

In this step, you should plan the entire project. You should understand the project well and identify the objectives and target goals. This will help you predict the project in the right manner.

Step Four: Measure and Manage the Workflow

If you want to improve the quality of your product and also increase the time you spend on creating the product, you should use a Kanban cycle.

Step Five: Using Scientific Methods for Optimization

You should use the Kanban board to create a new strategy. You should make some changes to the strategies that you use if you wish to improve the workflow. You can predict the changes in the workflow and the results if you use the Kanban task board. This will help you create your approach to complete the project.

Conclusion

Lean thinking and agile are a way of business and not just methods used to improve projects. Some businesses have dived into these approaches and use lean thinking and agile methods to improve businesses and processes. Both lean and agile project management methodologies will require a change to be made to the processes and also to the management and leadership. This is the only way the business will be open to new ideas and thoughts.

These new-age project management methodologies will make it easier for businesses to encourage their employees to identify new ways to improve processes, innovate and develop new processes, and identify issues and develop solutions to cater to those issues. This creates a sense of equality in the organization since every employee has the right to voice his or her opinion. It is important that you establish these processes in the firm right from the beginning to foster a sense of inclusion and innovation.

It is hard to identify the right method for your team at the first attempt. You will also take some time to get the hang of the process. Make sure that you understand the concepts well before you implement them in your organization or team. Remember that you cannot afford to make a mistake with implementation since that will cost the company a lot of money and time.

Thank you for purchasing this book. I hope you have gathered all the necessary information.

Resources

https://www.pmi.org/about/learn-about-pmi/what-is-project-management

https://zenkit.com/en/blog/7-popular-project-management-methodologies-and-what-theyre-best-suited-for/

https://activecollab.com/blog/project-management/project-manager-roles-and-responsibilities

https://www.smartsheet.com/content-center/best-practices/project-management/project-management-guide/how-choose-project-management-methodology

https://www.wrike.com/project-management-guide/faq/what-is-lean-project-management/

https://www.goskills.com/Lean-Six-Sigma

https://www.pmi.org/learning/library/lean-project-management-7364

https://www.projectengineer.net/how-to-perform-lean-project-management/

https://barryoreilly.com/what-is-lean-enterprise-and-why-it-matters/

https://www.whatislean.org/lean-enterprise/

https://leanstartup.co/build-dream-lean-team/

https://www.manufacturing.net/home/article/13193627/what-is-a-lean-team

https://www.revelx.co/blog/what-is-lean-analytics/

https://www.mendix.com/agile-framework/

https://www.softwareadvice.com/resources/agile-frameworks/

https://www.productplan.com/glossary/agile-framework/

https://blog.orangescrum.com/2018/11/step-by-step-guide-to-agile-project-management.html

https://www.cmswire.com/information-management/agile-vs-scrum-vs-kanban-weighing-the-differences/

https://blog.forecast.it/implementation-of-scrum-7-steps

https://hygger.io/blog/5-steps-of-kanban-implementation/

Made in United States
Troutdale, OR
02/22/2024

17900947R00084